INVESTING FOR SUSTAINABILITY
The Management of Mineral Wealth

INVESTING FOR SUSTAINABILITY
The Management of Mineral Wealth

by

Rögnvaldur Hannesson
The Norwegian School of Economics and
BusinessAdministration, Norway

KLUWER ACADEMIC PUBLISHERS
Boston / Dordrecht / London

Distributors for North, Central and South America:
Kluwer Academic Publishers
101 Philip Drive
Assinippi Park
Norwell, Massachusetts 02061 USA
Telephone (781) 871-6600
Fax (781) 681-9045
E-Mail < kluwer@wkap.com >

Distributors for all other countries:
Kluwer Academic Publishers Group
Distribution Centre
Post Office Box 322
3300 AH Dordrecht, THE NETHERLANDS
Telephone 31 78 6392 392
Fax 31 78 6546 474
E-Mail < services@wkap.nl >

 Electronic Services < http://www.wkap.nl >

Library of Congress Cataloging-in-Publication Data

Hannesson, Rögnvaldur, 1943-
  Investing for sustainability: the management of mineral wealth/ Rögnvaldur Hannesson
  p.cm.
  Includes bibliographical references.
  ISBN 0-7923-7294-8
    1.Alaska permanent fund corporation. 2. Alberta heritage savings trust fund. 3. Statens
    petroeumsfond. 4. Special funds 5. Environmental protection – finance. 6. Mineral
    industries – Environmental aspects – Case studies.

HJ3801 .H36 2001
332.67/252 21   00-067791

Printed on acid-free paper.   Printed in the United States of America

***The Publisher offers discounts on this book for course use and bulk purchases.
For further information, send email to <molly.taylor@wkap.com> .***

"It is a well-provisioned ship, this on which we sail through space. If the bread and beef above decks seem to grow scarce, we but open a hatch and there is a new supply, of which before we never dreamed. And very great command over the services of others comes to those who as the hatches are opened are permitted to say, 'This is mine!'"

Henry George (1879): *Progress and Poverty* (p. 243 in the edition from the Robert Schalkenbach Foundation, New York 1997).

TABLE OF CONTENTS

PREFACE

Experiencing at close quarters the emergence of the Norwegian petroleum industry over the last quarter century aroused my curiosity about resource windfalls and management of petroleum wealth. Norway is a small country but richly endowed with natural resources. One of these is, of course, the beauty of the landscape. I hasten to add that I'm expressing an unbiased opinion, because I was neither born nor raised in Norway. Then there are the hydro power and the fisheries. That notwithstanding, Norway used to be the poorest of the three Scandinavian cousins, but not by much.

Oil turned all that upside down. For several years Norway has had the highest GDP per capita of those three, although one might perhaps have expected that the difference would be greater, with Denmark having very little oil and gas and Sweden having found virtually none, despite considerable exploratory effort. But oil can be a mixed blessing.

On the whole, however, Norway has probably managed to deal better than most with the petroleum wealth. Nevertheless, one can ask if things could have been done better. One thing I find curiously absent from the Norwegian public debate is management of petroleum wealth in a long term perspective. While it is true that issues such as "what shall we do when the oil runs out?" pop up from time to time the perspective is one of preserving industry, or perhaps even industrial relic, rather than of investing oil revenues and making oil wealth permanent.

Some time in the 1980s I became aware of the Alaska Permanent Fund. It was also in the 1980s that the fund idea popped up in the Norwegian debate, but wealth preservation was never a major issue and giving people a direct share in the oil wealth through a dividend program like they do in Alaska was never mentioned. I began to familiarize myself with the Alaska Permanent Fund. Perhaps there was an example here for us to emulate. I also became aware of the Alberta Heritage Fund, and those two made an interesting comparison, as is discussed in the book.

Gradually the idea of writing a book about mineral wealth and investment funds emerged. During a sabbatical term at the University of Queensland, Australia, I had occasion to read some literature on mineral economics and investment funds. Back in Bergen it remained to pull it all together and see if it would justify a book. One justification is that not much has been written on the use of investment funds to preserve mineral wealth, so I believe that this book serves a purpose. It is not meant as a textbook, even if it contains some formal reasoning, but it is my hope that it will be useful as supplementary reading in mineral economics, macroeconomics, and development economics. Its relevance is, of course, greatest for countries which are rich in mineral resources, particularly oil and natural gas.

The book is not aimed solely at the academic economist. The few mathematical expressions there are should not be off-putting for the non-economist, and the arguments should be clear even if some formulas must be skipped for lack of the requisite mathematical background. My hope is that political scientists and policy makers will find this book useful, and that the interested layman will find it interesting. Besides mineral wealth the book also addresses the issue of sustainability, which for the last fifteen years or so has become a household word in wide circles.

*

I am grateful to several friends and colleagues who have read the manuscript and suggested improvements. Professor emeritus Anthony Scott of the University of British Columbia, one time teacher and mentor; professor Daniel Gordon of the University of Calgary; professor Thorvaldur Gylfason of the University of Iceland; Sarah Bibb, now with the National Marine Fisheries Service in Juneau, who knows more than most about the Permanent Fund; all of these have read parts of the manuscript and provided suggestions and corrections. Needless to say, none of them has any responsibility for errors that might remain. Nina Mollett, also with the National Marine Fisheries Service in Juneau, applied her editorial skills generously to parts of the manuscript. I also benefited greatly from interacting through the Internet with people in Alaska and Alberta whom I've never met and who certainly had no obligation to deal with queries from an unknown person on the other side of the globe. In general, the North American willingness to provide help and information deserves to be stressed and commended. I only hope that Norwegians and other Europeans in a similar position are equally forthcoming towards inquisitive researchers, but I did not need to put that to test on this occasion.

Bergen, October 2000

Rögnvaldur Hannesson

CHAPTER ONE
THE ELUSIVE CONCEPT OF SUSTAINABILITY

"... unless some auspicious expedient offer itself
and means speedily resolved upon for a future
store, one of the most glorious and considerable
bulwarks of this nation will within a few
centuries be nearly extinct. With all the
projected improvements in our internal
navigation, whence shall we procure supplies of
timber fifty years hence for the continuance of
our navy?"

J.D. Brown (1832): *Sylva Americana.*[1]

A tale is told of a poor farmer in Ireland who lived more than two
hundred years ago. There were those who were less lucky than he. His land
contained peat, a valuable fuel at the time. Even if the phrase had not yet
been coined, he was concerned about what we now call sustainable
development. He realized that the peat underneath his land would not last for
ever, but he was not just concerned about himself, his wife and his children;
he was also concerned about his grandchildren, his children's grandchildren,
and so on; in short he was concerned that each generation would get its fair
share of the wealth in the family's land. He figured out how much peat there
was likely to be underneath each square yard of land, and how much each
generation would need to make, well, not such a terribly great living but in
any case one on the happy side of subsistence. He divided his land into
squares for each generation and put down stones in the corners to mark them
off. He made a vow not to take more than his share and to maintain the stone
markings well visible and, when the time was ripe, persuaded his oldest son
to make a similar pledge.

The tradition was maintained in the family for a few generations.
But one day the digging of peat came to an end. The reason was not that
there was no more peat; the old man had been surprisingly correct in
estimating the amount of peat in the land, at least for the part that had been

---
[1] Quoted from Scott (1973), p. 244.

dug out. But there was no more market for peat; peat had been replaced by other and better fuels. The generation who suffered the collapse of the peat market cursed their ancestor as fiercely as he would have expected them to do if they had been deprived of their fair share of the peat. "Why did we heed the fantasies of that old fool", they said? "Why didn't we dig out the peat while it was still worth something?"[2]

We do not have to resort to fables for making the point that the value of non-renewable resource deposits can be eroded by technological change rather than by digging them out and disposing of them. In 1865 the famous English economist Stanley Jevons published a book entitled "The Coal Question". The main theme of that book was that England owed its leading position in the world to its cheap coal. Examples of the critical importance of coal were many and obvious. Coal was the fuel of the industrial revolution, by then about a hundred years old. Coal was what turned iron into steel and fueled the modern means of communication at the time; the steamship and the railroad train. A strong manufacturing industry and good communications were essential for a nation's prosperity and leadership in the world. Jevons saw the towering position of Britain as being threatened by the fact that coal was getting scarce in Britain; it was becoming necessary to dig deeper and deeper while competitors such as the United States and Germany still had plenty of cheap coal and could thus be expected to surpass Britain in the not too distant future. He put through the core of his message with a conviction characteristic of a Victorian gentleman: "The alternatives before us are simple. Our empire and race already comprise one fifth of the world's population; and by our plantation of new States, by our guardianship of the seas, by our penetrating commerce, and above all by the dissemination of our new arts, we stimulate the progress of mankind in a degree not to be measured. If we lavishly and boldly put forward in the creation of our riches, both material and intellectual, it is hard to overestimate the pitch of beneficial influence to which we may attain in the present. *But the maintenance of such a position is physically impossible. We have to make the momentous choice between brief but true greatness and longer continued mediocrity.*"[3]

Jevons devoted some space in his book to the dismissal of fanciful ideas to the effect that coal could be replaced by some other source of energy. A certain degree of irritation can be sensed between the lines as he painstakingly puts down the idea that coal could be replaced by petroleum. And other important means of communication than the railroad train and the steamship were not in sight. As he put it, "even if an aërial machine could be

---

[2] I am indebted to Chris Lightfoot, consultant to the Asian Development Bank, for this fable.
[3] Jevons (1906), pp. 459 - 460. Italics in the original.

propelled by some internal power from fifty to a hundred miles per hour it could not make head against a gale."[4]

It is easy, of course, more than a hundred years later and with the benefit of hindsight, to ridicule Jevons' predictions. That is not my purpose. Jevons was one of the best English economists of his day, and he made lasting contributions to economic theory. The economists of today, or anyone else for that matter, are not likely to make any better predictions than he did, except that they may perhaps be better aware of the speed and unpredictability of change and therefore less prone to venture out on the slippery slopes of making predictions. The example serves the purpose of illustrating how the value of resources is conditioned by our technology, how resources that may at one time seem invaluable may nevertheless become dispensable and worthless due to technological progress, and how difficult it is to predict technological change. The only thing I will hold against Jevons is that his fixation on how indispensable coal was is a little strange in the light of his discussion of the rise of the earth coal industry in England. Iron was traditionally made with charcoal. As a result, the forests of England were rapidly disappearing. This made many people concerned, some for what we would now call environmental reasons, while others were concerned that there would be too few big trees left to build ships. The story of how iron came to be made with earth coal instead of charcoal involved many failed experiments and entrepreneurs who went broke. It was well told by Jevons, and should have made him more aware of the unexpected turns of technological progress. Presumably there was a time when pundits predicted that iron could never be made with earth coal.

There are many other telling examples from earlier times when there was less contact between civilizations of how the value of resources is conditioned by technology and custom. The Spanish conquistadors searched with increasing frenzy the El Dorado, the land of gold, in what is now Colombia. Gold was the only metal the Indians of that part of the world knew how to extract and work with, but unfortunately it is not a very useful metal. One of the few useful utensils they made of gold was fish hooks, but otherwise they made ornaments more conspicuous for their size than their elaborate craftsmanship, something that might be due to the metal not being very precious and scarce for the Indians. For the Spanish gold was valuable as a medium of exchange and a store of value, due precisely to its scarcity.

Another example is the island of Nauru, to which we will return later, and other phosphate islands in the Pacific. Over the ages, ocean birds deposited their droppings on these isolated atolls in the vast Pacific, long before they became inhabited by humans. Over time these deposits hardened

---

[4] Jevons (1906), p. 169.

4

into rock, rich in phosphate due to its origin. Phosphate is, in turn, valuable fertilizer and especially so in the antipodean lands of New Zealand and Australia. The inhabitants of these phosphate islands were, in the beginning unknowingly, sitting on a resource that was extremely valuable for the farmers of New Zealand and Australia, but having little use for it themselves. Hence, and helped by the machinations of the colonial era, these deposits became a cheap source of fertilizer for the said countries and their motherland, Great Britain.[5] But a Norwegian does not have to go far afield to find examples of resources suddenly made valuable by technological progress while next to worthless for those ignorant of the new technology or not in a position to make use of it. In the late 1900s, in the early days of the age of electricity, two engineers from Bergen bought waterfall rights in the Hardangerfjord area from a local farmer for a few thousand kroner and sold them about ten years later for about 200,000 kroner each.[6] It tells us something about the importance of coal at that time that the waterfalls were often referred to as "white coal" by the panegyric enthusiasts of the new technology.

Sustainability: a nice idea in search of a meaning

Whatever sustainable development of our civilization means, it can hardly mean a process that repeats itself over and over again in an immutable fashion. Rather it would seem to mean a process that is able to continue by adapting to ever changing circumstances, which in part means overcoming problems of its own making. Somehow humanity has been able to sustain itself to this day and indeed multiply its numbers well beyond what once it was thought the earth would be able to support. This has been made possible by discovering new technologies, new sources of food and materials, and new ways to produce them. Looking back it is difficult to see human development in other terms than as a triumph over nature. We have learned the secrets of her laws and turned them to our own advantage, and overcome the constraints our ancestors faced as animals without tools and accumulated knowledge passed from generation to generation.

Against that background it is difficult indeed to understand much of the pessimism about the human condition preached by a number of prophets among us. True, there is one reason for pessimism that was not present before, or at any rate not to the same degree. The industrial revolution was a quantum leap in technology, and it changed human society beyond

[5] See Weeramantry (1992).
[6] Information on display in the industry museum in Tyssedal, Norway.

recognition. But the industrial revolution also meant a quantum leap in our dependence on non-renewable resources. As Jevons was so well aware, the industrial revolution was fueled by coal, a non-renewable resource consisting of stored solar energy flowing millions of years ago and not retrievable once the coal has been burnt. Furthermore, the prime material of the industrial revolution for making things was iron, and iron ore is another non-renewable resource. Since non-renewable resources are by definition limited and cannot be exploited indefinitely, any civilization that makes itself dependent on them would seem to be in great peril, coming to an end when the resources run out. But things are not as straightforward as they may seem. First, next to nothing is known about how much there is of non-renewable resources of various kinds. The figures on remaining reserves of such resources that one finds in statistical publications are not the result of a complete inventory of what lies hidden in the earth's crust but simply an inventory of deposits the mining companies have found it worthwhile to look for. The mining companies are not interested in spending money on finding more than is needed for a comfortable planning horizon for their operations. This is why the reserves of oil in the world have corresponded to 30 – 40 years of contemporary production ever since the end of the Second World War, even if the annual production of oil has increased tenfold over that same period.[7]

Secondly, as the development since Jevons' days has shown, it has been possible to replace one kind of non-renewable resource with another. Britain has indeed slipped from its leading role in the world as he feared would happen, but not because it ran out of coal. Coal has to a large extent, but certainly not entirely, been replaced by oil and gas. Copper and iron have, again to some extent, been replaced with other metals like aluminum. Aha, you might say, what does that help; these are also non-renewable resources. Indeed, this might only succeed in further delaying the Doomsday when all our minerals are exhausted, although that prolongation should not, perhaps, be considered trivial. What will have to happen, however, if we are to avoid the threat ultimately posed by our reliance on non-renewable resources, is that non-renewable sources of energy be replaced by renewable sources of energy, and non-renewable materials by renewable materials derived from animals and plants or what with today's technology is useless and abundant rock. How this will happen, when it will happen, and whether it will happen at all none of us can tell, but it is already now possible to

[7] In 1937 world oil reserves were estimated at 24 billion barrels. Oil production in 1938 was 5.4 million barrels per day, so the reserves amounted to not much more than twelve years of production. Small wonder, perhaps, that in 1937 the journal Petroleum Economist published an article entitled "Is there Enough Oil?" (On this, see "Fifty Years of the Oil Industry," Petroleum Economist, September 1983, pp. 326 – 327.) In 1998 world oil reserves amounted to more than forty years of production (BP Amoco Statistical Review of World Energy).

harness solar energy, and wind and wave energy as well, albeit at a high cost. As the truism has it, predicting events is difficult, especially those that have not yet happened, and we are probably no better at doing it than Jevons.

Substitution of renewable resources for non-renewable sources of energy and materials will never happen, however, unless we devote some of the riches we derive from our present exploitation of non-renewable resources to acquiring the skills that will make such transition possible. This brings us to the theme of investing for sustainability, the setting aside of temporary and unsustainable gains for purposes that will make these gains permanent instead of transient. This involves investment in knowledge, experimental equipment, anything that would enable us to replace non-renewable resources with renewable ones, as the need arises. These are truly momentous problems, concerning what we might call the grand dynamics of humanity. Nevertheless there really is very little that can be said about them with any confidence and precision. There is some reason to believe that the transition away from reliance on non-renewable resources will happen more or less automatically; as resources become scarce they also become expensive, so that finding substitutes becomes increasingly rewarding. But such generalities are about all that it is possible to say about these problems.

Preservation of mineral wealth

These grand dynamics, about which so little can be said with firm confidence, are not the main topic of this book. While it deals with the transformation of non-renewable resource wealth to renewable wealth it does so in much less dramatic terms. The main problem to be discussed is that of a country, or a province, rich in mineral deposits that could reasonably be expected to run out in the not too distant future. In global terms the mineral involved may be abundant enough, on any reasonable time scale; there may be undiscovered deposits elsewhere, perhaps of a poorer quality, but nonetheless the future of industrial civilization need not be at risk if one single country runs out of minerals. But a country facing dwindling mineral deposits within its territory, or at any rate dwindling deposits of good quality, also faces the problem of becoming poorer unless it has managed to transform its non-renewable resource wealth into permanent, renewable wealth. Unlike fish in the sea or trees in a forest, mineral deposits are non-renewable; they do not grow in the ground. Hence it would appear that as the mineral is being dug out the mineral wealth is being diminished, unless it is transformed into some form of renewable wealth. Things can be a bit more complicated than that, however, because the mineral remaining in the ground may rise in price to such an extent that the mineral wealth increases despite physically dwindling reserves. We shall discuss this probably unlikely case a bit more explicitly in the next chapter.

Barring a price increase of a magnitude discussed in the previous paragraph, a country that is fortunate enough to have profitable deposits of minerals within its borders will not be any richer in the long run if it neglects to invest its mineral income in renewable wealth of some kind. It would be able to ride high for a while and enjoy an unsustainable standard of living while digging up its minerals, but once they're gone they're gone, never to be retrieved. In fact it could be argued that a country like that would even be worse off because of its mineral wealth; adjusting one's standard of living downwards can be a painful process, for individuals and societies alike.

In principle it is straightforward how mineral wealth can be preserved; it must be invested in something that permanently increases the mineral owner's command over goods and services. If the reader finds this phrase convoluted it is so deliberately, because it is in fact not quite straightforward what this entails; it depends on who the mineral owner is and how that entity's choices are constrained. If the mineral owner is a nation which does not trade with other nations, the only way in which that nation can increase its wealth permanently is by raising its productive capacity through investment in production equipment like machines and buildings, in infrastructure like roads or whatever smooths the geographical flow of goods and services, or in human capital like knowledge and health which makes the affected individuals more productive. But in this day and age of globalization, nations that do not trade with other nations are hard to find. A nation which is rich in minerals and trades with other nations can increase its wealth by increasing its stock of foreign financial assets, which are simply claims on the fruits of other people's work. If the mineral owner is an individual or a firm it is even more obvious that the wealth of the mineral owner would ordinarily be raised by increasing the holdings of financial assets rather than the productive capacity of that individual or even that of the firm. Ultimately those increased holdings of financial assets must be backed by a corresponding rise in the productive capacity somewhere, unless the wealth of the mineral owner is being augmented at somebody else's expense. For the world as a whole, however, the metaphor of a nation that does not trade with other nations is fully valid. If the global mineral wealth is to be preserved, it must be invested in productive capacity, be it real capital or human capital.

How have nations chosen to use their mineral wealth? What is it that entices them to preserve it rather than to waste it by consuming the income from their minerals as soon as they have been dug out of the ground? Conducting case studies of this is in fact less easy than it might appear. First, what is the mineral income to be invested? "Income" is in fact something of

8

a misnomer, since we are dealing here with wealth rather than income, but we postpone the discussion of this until the next chapter where the concept of rent will be explained. How much of the mineral income should be invested depends on a number of factors, to be discussed in Chapter Three. And, finally, how much have nations invested of their mineral incomes, compared to what they should have? Again this is not easy to establish, because what we need to look at is the counterfactual question of what they would have done if there had been no mineral income.

Studies of economic growth have indicated that abundance of natural resources hinders rather than helps economic growth.[8] A study by the World Bank of the effects of the windfall gains made by selected oil producing countries in the wake of the two oil price rises in the 1970s indicated that the countries studied had made little or no gains from the windfall.[9] The reason was not always that too little of the oil money was saved and invested; taken together the countries saved just about as much of it as a sustainable use would prescribe.[10] But most of the investment was undertaken by the government and turned out to be unproductive; steel mills or aluminum plants that never worked to capacity or never made a profit for other reasons, and investments in infrastructure or education that did not result in significant productivity gains. In other cases weak governments placated an impatient electorate by using the oil windfall for non-sustainable increases in consumption rather than investment. Venezuela did both, and did badly despite setting out with good intentions. The president of Venezuela at the time, Carlos Andrés Pérez, wanted to use the oil money to make Venezuela a modern, industrially developed country. In 1979 he told a visiting American scholar "one day you Americans will be driving cars with bumpers made from our bauxite, our aluminum, and our labor. And we will be a developed country like you".[11] But Juan Pablo Pérez Alfonzo, the founder of OPEC, had misgivings. "Ten years from now, twenty years from now ... oil will bring us ruin." And he referred to oil as the "the devil's excrement."[12]

But industrially underdeveloped countries like Venezuela and Algeria are not the only ones that appear to have saved very little of their windfall gains from oil and gas. Both the United Kingdom and the Netherlands appear to have saved very little or nothing of the windfall gains that the energy crises of the 1970s brought them. The emergence of the oil and gas sector in the United Kingdom does not appear to have increased total

---

[8] See Gylfason (1999) and Sachs and Warner (1995)
[9] Gelb and associates (1988). The countries studied were Algeria, Ecuador, Venezuela, Trinidad and Tobago, Indonesia, and Nigeria.
[10] According to the Hicksian permanent income rule, to be discussed in Chapter Three.
[11] Karl (1997), p. 4.
[12] Karl (1997), pp. xv and 242.

investments by the United Kingdom at all; if anything, the opposite has occurred.

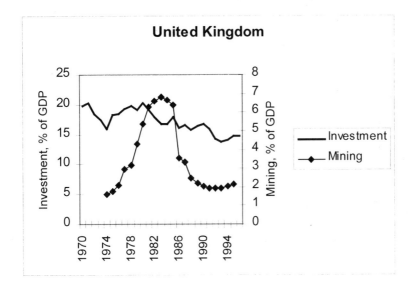

Figure 1.1. United Kingdom: Total investment, and mining and quarrying, as share of GDP. Source: OECD: *National Accounts*, 1998.

Figures 1.1 – 1.3 show the total investment in the three main energy producing countries in Europe, the United Kingdom, the Netherlands, and Norway, as share of their gross domestic product (GDP).[13] Also shown is the contribution of oil and gas (in the UK, the production of minerals) to GDP (note the different scales on the charts). In the United Kingdom gas production began on a small scale in the 1960s, but the total production of oil and gas received a big boost in the mid-1970s due to the development of the oil fields in the North Sea. The share of mining and quarrying increased from less than two percent of GDP in 1974 and 1975 to nearly seven percent in the mid-1980s, but fell thereafter to about two percent in the 1990s. This increase was due in part to the rising prices of oil and gas up to 1980, and in part to the increase in production of oil and gas on the British continental shelf. British investments have nevertheless fallen almost without interruption from about 20 percent of GDP around 1980 to about 15 percent

---

[13] That is, the sum of gross investment in real capital in the domestic economy and the current account surplus, the latter corresponding to net investment abroad.

10

in the early 1990s, indicating that little or nothing of Britain's oil and gas wealth has been transformed into renewable wealth.[14]

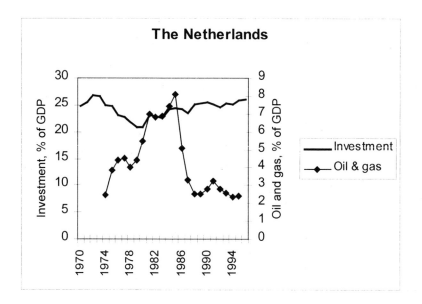

Figure 1.2. The Netherlands: Total investment, and oil and gas, as share of GDP. Source: OECD: *National Accounts*, 1998.

    The picture is not very different for the Netherlands (Figure 1.2). Gas production started in the Netherlands in the 1960s but did not for many years amount to a large share of GDP, due to low prices of energy in general and an aggressive gas pricing policy on behalf of the Dutch government in particular.[15] The contribution of oil and gas to the GDP rose from less than three percent in 1974 to eight percent in 1985 and then fell back to between two and three percent. Most of the rise is likely to have been an increase in pure profit, or rent (a concept to be explained in the next chapter), but very little or none of this was saved and transformed into renewable wealth.[16] Dutch investments, domestic and foreign, fell in the 1970s, then rose slightly in the early 1980s as the gas price windfall was at its high point, and in fact went on rising as the value of the gas production fell, but was no higher in the early 1990s than it had been in the early 1970s, before the gas price windfall.

---

[14] Forsyth, in a much earlier study, reaches a similar conclusion (Forsyth, 1986).
[15] On this, see Kremers (1986).
[16] Kremers (1986) reached a similar conclusion in the mid-1980s.

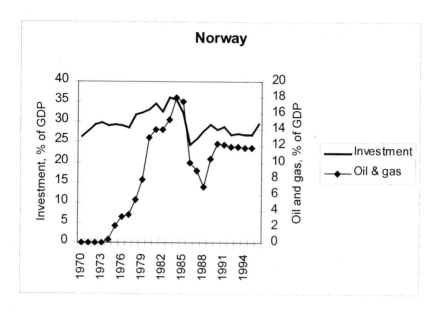

Figure 1.3. Norway: Total investment, and oil and gas, as share of GDP. Source: OECD: *National Accounts*, 1998.

Norway is a bit of an exception (Figure 1.3). The oil and gas sector was negligible until the mid-1970s and then grew rapidly. Its production value dipped in the 1980s with the collapse of the oil price but the sector nevertheless remains large, its share of GDP in the early 1990s being about 12 percent. Norwegian investments, always a high share of GDP, increased from 26 percent of GDP to a maximum of 35 percent which coincided with the maximum of the oil and gas production value in 1985, and then fell back to about 25 percent as the oil price collapsed. Lately however, as we shall discuss in a later chapter on Norway, the investments have again increased as the value of oil and gas production has soared. It thus appears that Norway has transformed a substantial amount of its oil wealth into renewable wealth, unlike the other two. In this it has undoubtedly been helped by being a small country with a limited capacity to absorb large export revenues, in contrast with the other two, particularly the United Kingdom.

Investment funds

One thing both the Netherlands and the United Kingdom, and for a long time Norway as well, did not do was to channel some of their oil and gas revenues into an investment fund. Would this have helped them save some of their oil and gas wealth? Venezuela tried this, but the money stopped flowing into the fund after only a few years. That notwithstanding, we may ask whether such a strategy would facilitate a sustainable management of mineral wealth. For one thing, this would make the mineral incomes easier to identify. How successfully the mineral incomes would be invested would depend on the incentive structure characterizing the investment fund; it is not difficult to envisage investment funds which, despite their name, would be little other than black holes to throw money into, and examples are probably not difficult to find in the real world either. Last but not least, a sufficient amount of money would have to be allocated to the fund, as the Venezuelan example reminds us of.

There are a number of countries which have used, or tried to use, investment funds for preserving their mineral wealth. One is Nauru, which has channeled some of its phosphate incomes into several investment funds. Another is Kuwait, which built up investment funds from its oil revenues explicitly for the benefit of future generations, but was forced to use some of that money prematurely because of the Iraqi invasion in 1990. There are other examples of such funds financed from oil in the Middle East.[17] Brunei, a small country on the island of Borneo and rich in oil, has also set aside some of its oil revenues in an investment fund, but the success of that operation has recently been called into question.[18] In North America there are several examples of investment funds having been built up from mineral revenues; such funds exist in New Mexico, Wyoming and Montana.[19] The biggest ones, however, in North America are the Alaska Permanent Fund and the Alberta Heritage Fund, both of which have been built from petroleum revenues. In Norway an investment fund financed from petroleum revenues was set up in the 1990s.

In this book we shall discuss three petroleum funds, the Alaska Permanent Fund, the Alberta Heritage Fund, and the Norwegian Petroleum Fund. Alaska, Alberta and Norway have many things in common, despite obvious differences. Alaska and Alberta are parts of a much larger, federal state while Norway is a sovereign state. The difference is not as fundamental, however, as it may appear. All three have large petroleum

---

[17] Some of these funds are discussed in Stauffer (1988).

[18] The Brunei Investment Agency was run by one of the Sultan's brothers and is believed to have diminished in value from 110 billion US dollars to about 40 billion over a few years in the 1990s. The Sultan released his brother from his responsibilities in early 2000 and started court proceedings against him, but gave him a subsistence allowance of 300,000 dollars per month to support his four wives and 35 children, according to the Financial Times of March 28, 2000. The private wealth of the prince was substantial; he was believed to own 17 airplanes and 2000 cars, according to the same source.

[19] See Pretes and Robinson (1988).

resources relative to their populations. The state of Alaska and the province of Alberta have considerable autonomy and get most of the tax revenue from petroleum exploitation within their borders. Both of them have their own legislative assemblies and provincial versus state governments answerable to their electorates. In terms of population Norway is not much larger than Alberta, there are about four and a half million inhabitants in Norway against three in Alberta, but Alaska is much the smallest, with only about six hundred thousand inhabitants. All three may thus be characterized as small, regional economies, although the Norwegian economy is not as well integrated with the European Union as Alberta and Alaska are with the rest of the federal states to which they belong.[20] As we shall see, the experiences of these three entities with respect to their petroleum investment funds are different, as are the constitutional and legislative arrangements in which they are embedded, providing different incentives with respect to investment of mineral tax revenues and preservation of mineral wealth. Here there may be lessons to be learned for other mineral rich countries or provinces which might want to take this route to preserving their mineral wealth.

A brief summary of the book

In the next two chapters we discuss some fundamental issues pertaining to mineral wealth and investment funds. The next chapter explains the concept of mineral rent, but mineral wealth is nothing but the present value of mineral rents. Three types of mineral rents may be identified; differential rent, pure scarcity rent, and market power rent, although the last type really is just a perturbation of the time profile of rents. The famous Hotelling theory of the time profile of rents is briefly reviewed and contrasted with the development of mineral prices over the last century, a comparison that is not supportive of the theory. The exposition is a bit formal, but readers who have difficulties in following the half a dozen or so formulas should nevertheless be able to follow the arguments in the text. The chapter concludes with some musings on who are entitled to mineral rents.

Chapter Three starts with applying the Hicksian concept of income to mineral extraction. The idea is simple; how much can a country consume out of its current mineral revenues without impoverishing itself in the long run?

---

[20] Norway is not, in fact, a member of the European Union. It is, however, together with Iceland and Lichtenstein, a member of an entity called the European Economic Area, which is a forum of cooperation with the European Union. This means in effect a common market for most goods and services, and a common labor market. Furthermore, the regulations issued by the European Commission in Brussels apply to Norway as well, with certain exceptions. Hence the integration with the Union is far reaching.

The exposition is supported by some simple models and formulas which, it is hoped, even the mathematically non-inclined reader will find useful and clarifying. It is demonstrated how the amount that needs to be saved annually in order to ensure that mineral wealth will be preserved depends on the rate of return and the time profile of the mineral revenues. A long time horizon and a high rate of return lighten the savings burden considerably. Saving a constant share of a declining revenue is shown not to be particularly sensible, appealing as such a rule of thumb might be from a political point of view. The chapter also discusses who could be entrusted with saving rents, governments or private companies. There are political and distributional reasons why there is a role for governments here, but governments may fail to serve their citizens for reasons of short-sightedness that ultimately may lie with the electorate, or for reasons of corruption. Other questions discussed in this chapter pertain to the design of investment funds used for preserving mineral wealth. What should such funds invest in, public goods or projects with an identifiable financial return, and should they invest at home or abroad? And, finally, what should be done with their return? It is argued that giving individuals a stake in such funds would be likely to promote preservation of mineral wealth.

The following four chapters discuss the experience of four mineral rent investment funds; the Nauru phosphate funds, the Alaska Permanent Fund, the Alberta Heritage Fund, and the Norwegian Petroleum Fund. All of these are small, open economies with a large and, in the case of Nauru, utterly dominant mineral resource sector. All four have invested a substantial share of their mineral revenues in these funds, but it is unclear whether in all cases a long term preservation of mineral wealth has been the major goal or whether the fund was meant just to shift the consumption of mineral wealth over time. All four have not met with an equal measure of success. Nauru's phosphate funds have largely failed. The phosphate mining in Nauru is a stark example of non-sustainable activity; it has devastated the island, and insufficient provision has been made for the continued support of the standard of living that the phosphate mining has made possible, an activity that is now coming to an end. Chapter Four deals with the case of Nauru.

The most successful of the four funds is the Alaska Permanent Fund. It is, perhaps, unmistakenly American. Its existence owes much to the efforts of an unconventional governor of the State of Alaska, elected in opposition to much of the political establishment. One of his major goals was to keep the oil money of the state out of the hands of spendthrift politicians and to make sure it was used for the benefit of the ordinary Alaskan, in the present and the future. As the name indicates, the fund is meant to preserve the state's mineral wealth for the indefinite future, and the return on the fund is distributed among the people of Alaska in the form of a dividend. The dividend program appears to have made the ordinary Alaskan highly interested in the fund and its management but, at the same time, the state of Alaska has gotten itself entangled in fiscal problems. The long term solution to these problems is either to reintroduce the income tax or cut into the people's dividend. Neither is likely to be popular. The Alaska Permanent Fund and the fiscal problems of Alaska are discussed in Chapter Five.

The nearby Canadian province of Alberta is also rich in oil and gas. The windfall gains of the first energy crisis prompted the Alberta legislature to channel some of the oil and gas revenues into an investment fund called the Heritage Fund. This fund never became as large or as successful as the Alaska Permanent Fund, and was in effect rendered fictitious as the province began to accumulate debt in the 1980s. The Heritage Fund was never a "popular" institution the way the Alaska Permanent Fund is; it has no dividend program, it was managed by the Alberta Treasury and overseen by the Alberta Parliament, and its returns were incorporated into the general government budget. Public awareness of and interest in the Heritage Fund seems low. The Alberta Heritage Fund is discussed in Chapter Six.

Norway's national wealth was immensely increased by the discovery of oil and gas on the North Sea continental shelf and the increase in the oil price in the 1970s. Only lately, however, has the Norwegian Parliament decided to set up an investment fund to manage this wealth. It is unclear, however, whether the purpose of this fund is to preserve the wealth or just shift its consumption over time. The Norwegian Petroleum Fund is very much in the European parliamentary tradition; the annual deposits or withdrawals are at the discretion of the parliamentary majority and there is no dividend program or any other mechanism which gives the individual or household a direct share in the return on the fund. The history of the fund is too short to draw any firm conclusions, but the omens are not all good. The Norwegian Fund is discussed in Chapter Seven.

The final chapter draws conclusions from the experience of the three petroleum funds. Are such funds useful for making mineral wealth permanent or are they just accounting fictions? What are the incentive mechanisms needed to make such funds successful? Are the experiences in places like Alaska, Alberta and Norway of any interest in other parts of the world where circumstances may be very different?

CHAPTER TWO
MINERAL WEALTH: WHAT IT IS

"No longer will miners be ordered to "pick the
eyes" of underground seams and veins, since it
will pay to work out a lower average grade than
they had before."

Anthony Scott (1955): *Natural Resources. The
Economics of Conservation.*[21]

Often, the extraction of non-renewable resources is extremely
profitable. Diamonds and oil are prime examples. Revenues from diamonds
are believed to be the main source of income for the war machine of Unita, a
rebel movement fighting the government of Angola. Diamonds are also the
main factor behind the economic growth in Botswana, which for the last two
decades has had growth rates comparable to the Tiger economies of
Southeast Asia.[22] The profit per barrel of oil varies greatly from place to
place, due to differences in costs. The cost of production probably is about
two to five US dollars per barrel in Saudi Arabia, but in the North Sea it can
be twice as high or more. The price of oil has for the most part been well
above that but volatile; over the two-year period mid-1997 to mid-1999 the
price of oil fell from about 20 US dollars per barrel to a low point of about 10
dollars at the end of 1998, and rose again to about 22 dollars just half a year
later. In 2000 it periodically went beyond 30 dollars.

The difference in profits from place to place is typical not just for oil
but for non-renewable resources in general. The main reason is differences in
costs of production and transportation; while product items of a similar
quality fetch similar prices in the marketplace there are vast differences in the
richness of deposits and the ease with which they can be accessed. Drilling
for oil in the sands of Arabia is much cheaper than in the North Sea, where it
is sometimes necessary to construct platforms of concrete that would dwarf
the Eiffel Tower. A barrel of oil can be produced more cheaply from a large

---

[21] Page 21 in the 1973 edition.
[22] See Auty and Mikesell (1998).

oil field than a small one, and a tonne of copper more cheaply from rich ore than a poor one. This difference between the product price and all necessary costs of production is called rent. Just as fertile land or an inner city studio can be rented out at a higher price than poor land or a flat in the suburbs, a profitable mine could be rented out, or sold, at a higher price than a less profitable one. The difference in profitability is what accounts for the difference in the rents, or the prices, of mines; the amount a lessee would be willing to pay for renting a mine would be capped by the annual profit to be made from the mine, and the upper limit of the purchase price would be the present value of the future profits of the mine.[23] Of course the lessee, or the buyer, would prefer to pay less, but if there is sufficient competition among potential lessees or buyers the owner of a mine might come close to getting the entire profit of the mine as rental income or selling price.

These differences in rent are inherent in the physical and geographical differences between deposits of non-renewable resources and do not disappear even if markets are competitive. Increased competition may lower the price of the resource product and erode the rents of extraction—the market power of the Organization of Petroleum Exporting Countries (OPEC) was eroded by increased production of non-member countries in the 1980s— but the difference between the good mine and the poor mine will always remain. This is one of the characteristics that make the natural resource industries different from manufacturing industries. Car assembly plants, using the same technology and management methods, can be built anywhere. Over time they have been increasingly moved to countries where the wage cost is low, in response to the difference in profitability between car assembly plants in high wage and low wage countries. But as this process continues, together with the relocation of other manufacturing industries for the same reason, the difference in profitability between car assembly plants will disappear, partly because some will have been moved away from high wage countries and partly because the wage level in countries where the wages used to be low will have risen in response to industrial growth in these countries. There is no such mechanism at work in the natural resource industries; the poor mine cannot be moved, and the good mine will always be the good one and the poor one poor, unless some technological changes occur that erase the difference between what is a good and what is a poor mine. Hence, while investment and relocation of manufacturing firms in response to

---

[23] "Present value" is the value now of a payment occurring some time in the future. The present value of 100 dollars in a year's time is less than 100 dollars, because less than a 100 dollars could be put into a bank account now or invested in bonds to grow to 100 dollars a year from now. Technically, the present value of 100 dollars t years from now is $1/(1+r)^t$, where r is the rate of interest.

differences in profitability will erode wage differences the resource rent differentials in natural resource industries will remain.

It is the mineral rent, or the present value of a stream of mineral rents over the period for which they last, which constitutes mineral wealth. Note what, at this point, appears to be its finiteness; the total rent of a mineral deposit is equal to the rent per unit times the size of the deposit, and the latter is in principle finite if unknown. What makes this story a bit more complicated is that the rent per unit of a mineral could change over time. We address that problem below, but first a few more remarks on what determines the amount of rent per unit of a mineral at a point in time.

The explanation just given for the rent is one that goes under the name of differential rent, i.e., some deposits are more productive than others, or are more easily accessible, and hence produce more rent per unit than others. But there are two additional sources of rent. One is pure scarcity. Going back to the discussion of differential rent, the least advantageous deposit would earn no rent. Or would it? It all depends on whether it would be possible to satisfy the demand for the mineral from all existing deposits. For a long time to come this would, of course, be possible but not necessarily for the indefinite future. In the long term all existing deposits of a mineral, those known and those yet unknown, need not be sufficient to satisfy all demand. The marginal "deposit" would then be an alternative but more costly source of the service derived from the mineral, such as energy produced from renewable sources providing the alternative to fossil fuels. The mineral, in this case the fossil fuels, would earn some pure scarcity rent, due to their cost advantage vis-à-vis the alternative source. Needless to say, this scarcity rent cannot be known precisely, but to the extent it is relevant it could be forecast and its present value calculated from its value when the fossil fuels run out and the time lapse until this occurs. We return to this below when we consider how the rent of a mineral may change over time.

In addition to differential rent and pure scarcity rent there is yet a third component due to market power. If mineral producers are able to coordinate their sales and withhold some mineral from the market they will be able to extract a higher price than otherwise. In at least three cases this has been done with some success. One is OPEC's control of the oil market, which in certain periods at least has been effective enough to affect the price of oil; in 1997-98 during the Asian crisis the production of some OPEC members in excess of their quotas contributed to the fall of the oil price, while tightening of and adherence to production quotas took the price back to its previous level by mid-year 1999. The second case is the tin cartel. Most of the tin supplies of the world used to come from only four countries; Bolivia, Indonesia, Malaysia and Thailand, and the price of tin was controlled by successive International Tin Agreements beginning in 1956 but with predecessors dating back to the 1920s. In the 1980s Brazil and China established themselves as low cost producers and the tin cartel collapsed, with the price falling from about 18 US dollars per kg to less than 10 dollars, and since then it has declined further.[24] The third case is De Beers' control of the diamond market; this South African company sells about 70 percent of all the world's diamonds, even those coming from as far as Russia.

It can be argued, however, that the market power rent only amounts to a different time profile of rents but is not really an additional component of mineral rent. The owners of a non-renewable resource have nothing to gain from permanently withholding their resource deposits. This contrasts sharply with the classical monopolist who is able to permanently raise his profits by permanently reducing the flow of output or sales. The only degree of freedom available for those who might have gained monopoly power over the supply of a non-renewable resource is choosing a different time profile for the extraction than otherwise would prevail, which amounts to choosing a different time profile for the rents.

The time profile of rents: Hotelling's intertemporal price theory

As already stated, the rent per unit of a mineral may change over time. The rent per unit is equal to the difference between the price of the mineral and the cost per unit. This difference changes over time because both the price and the cost per unit change over time. There is a theory of mineral rents which predicts that the rent per unit will rise over time. This theory is often named after the American economist Harold Hotelling who early on stated it succinctly (Hotelling, 1931), but the idea can be traced back at least to Gray (1914). The rationale of this theory is as follows. The owner of a mineral deposit is faced with a choice between two basic alternatives, he can let his deposit remain in the ground or he can dig it up and sell the mineral in the marketplace. What he decides to do depends on by how much he expects the value of his resource deposit in the ground to grow. If it does not grow at all he will be much better off by digging up his mineral and selling it, investing the money in something that gives a positive return. So, unless the value of a mineral remaining in the ground grows at a rate comparable to the rate of return on investing the profit from digging it up, it will all be dug up and dumped on the market. But that, the story goes, would depress the price so that it could be expected to grow in the future. Hence we get an intertemporal market equilibrium with a continuously rising rent.

---

[24] See "Commodities in the 20th Century", a special feature in "Global Commodity Markets", January 2000, published by the World Bank.

Formally, if we denote the price of the mineral in period t by P_t, the unit cost by C_t, and the rate of interest by r, we get[25]

$$(1+r)[P_t - C_t] = P_{t+1} - C_{t+1}$$

The left hand side of this expression shows the value the mineral owner will have after one period if he extracts one unit of the mineral in period t and invests the rent in a financial market yielding the rate of interest r. The right hand side shows the value he will have in period $t+1$ if he lets one unit of the mineral remain in the ground. Clearly, if he is indifferent between the two alternatives, both sides must be equal. Note that the rent may rise over time either because the price rises or because the cost falls. One implication of a rising price over time, in a world with deposits of variable quality, is that mineral deposits which are not profitable with today's prices and technology could become profitable later.

Using the above formula recursively, we find that the rent of a mineral unit in period t will be

$$P_t - C_t = \frac{P_T - C_T}{(1+r)^{(T-t)}}$$

where $P_T - C_T$ is the rent of the last unit extracted, which occurs in period T. This rent would be equal to the cost advantage of the mineral at the moment the last unit is dug up and a transition has to be made to a substitute technology. The rent in period t would thus be equal to the discounted value of the rent in the last period of extraction. Another interpretation is that the rent will rise at a rate equal to the rate of interest. In the above formulation the rate of interest has been assumed to be constant. Nothing would change in principle if we allowed it to change over time; the formula would become a bit more messy, the rent would still rise at a rate equal to the rate of interest, but that rate would not be constant.

---

[25] This formula is not valid for the case where C depends on the remaining stock of the resource, but only for the case where C is either constant or predicted to fall over time due to technological progress. Stock-dependent costs make the intertemporal optimization problem more complicated and add little insight relevant to the present context. Note, further, that the single operator is assumed to take the price in each period as given and not to be able to affect it appreciably by his output decisions, but the difference between the rent in any two periods must be expected to change so as to make each operator indifferent between producing a marginal quantity now and at some time in the future. If this were not satisfied, all price taking operators would withdraw their production or increase it, and the sum of their decisions would affect prices.

An implication of Hotelling's theory is that the increase in the rent will compensate for the fact that some of the mineral deposit has been dug out; there will be less mineral left, but whatever is left will have become more valuable. It is even possible that the increase in the rent will overcompensate for the depletion of the mineral deposit, so that the total value of the mineral left in the ground will be increasing. To see this, it is easiest to use the continuous time variant of the above formula, with no extraction cost. In this version, the price evolves over time according to the formula

$$P_t = P_0 e^{rt}$$

The price determines how much will be sold of the mineral at each point in time. Let the quantity sold depend inversely on the price, as follows:

$$Q_t = AP_t^{-b}$$

where Q_t is the quantity demanded at time t, and A and b are parameters. The total quantity extracted cannot exceed the deposits available at time 0, which we denote by S_0. The total quantity extracted is the extraction in each period summed over all periods, or, in continuous time, Q_t integrated over time from point zero to point T when the extraction ends:

$$\int_0^T AP_t^{-b} dt = \int_0^T A(P_0 e^{rt})^{-b} dt = \frac{AP_0^{-b}}{rb}\left(1 - e^{-rbT}\right) = S_0$$

In the absence of an upper limit to the price (no substitute technology) the extraction will go on for ever, with less and less being extracted as time goes on, so T will be infinite. The value of the integral is nevertheless finite, so with given parameters A, b and r, the "planning problem" is to find the appropriate value of the initial price P_0, so that the left hand side will be equal to the right hand side (S_0) and the mineral will not be extracted prematurely. With T approaching infinity, we get

$$S_0 = \frac{AP_0^{-b}}{rb}$$

The value of the remaining reserves at any time point t is $P_t S_t$. Its rate of change is

$$\frac{dP}{dt}S + \frac{dS}{dt}P = rP_tS_t - Q_tP_t = P_t(rS_t - AP_t^{-b})$$

The sign of the expression on the left side depends on the sign of the expression within parentheses on the right side. Looking specifically at the case where the time horizon approaches infinity, as in the expression for S_0 above, we get for the expression in parentheses:

$$rS_t - AP_t^{-b} = AP_t^{-b}\left(\frac{1}{b} - 1\right)$$

So, if $b < 1$ (inelastic demand), the expression is positive and the value of remaining reserves rises over time. The implication of this is stark. A country that owns a share of the total mineral reserves and extracts in a way that keeps that share constant will become richer and richer over time, even if it consumes all the wealth it extracts and does not invest any of it; the "investment" that consists of leaving some of the mineral lying idly in the ground will be sufficient to keep the country's wealth increasing, because of the rising price of the mineral. The case of inelastic demand ($b < 1$) is perhaps none too likely, but it is worthwhile to note how a rising price could go far in compensating for physically dwindling reserves.

This story does not, however, match reality too well. The main problem is not that the mathematical example is a bit special, but other, more fundamental assumptions, are a little too unrealistic. It is not costless to dig minerals out of the ground. The rate at which it should be done depends on an analysis of how much is there and how the extraction should be spread over time to recover the usually very high capital cost of getting the mineral out of the ground. And, as has already been stressed, the question is not one of disposing of given and known reserves over some period of time, it is one of spending money on finding new reserves and then to invest as appropriate to mine them. The choice that a mineral company faces is more like this: how long can we stay in business with the reserves we have? How much should we spend on finding more? And, having found more, how much should we invest in an extraction facility in order to extract our reserves over time in the best possible way? Even with a mineral price that remains constant over time the extraction would be spread over time, simply because it would not make any sense to invest in production facilities as required for getting all the mineral out at once.

24

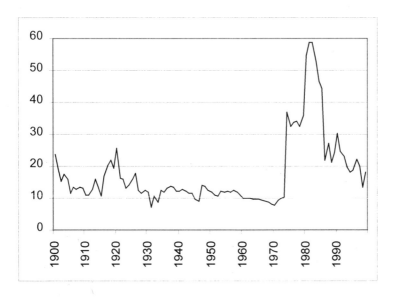

Figure 2.1. The price of oil 1900-99, in 1999-US dollars per barrel. Sources: *BP-Amoco Statistical Review of World Energy, Statistical Abstracts of the United States* (Consumer Price Index).

How mineral prices have evolved

Looking at mineral prices in a long time perspective, there is little evidence that they have been rising over time at a rate comparable to the rate of return on investment in productive assets, financial or real. Figure 2.1 shows the price of oil in 1999-US dollars over the last century. The price happened to be just about the same at the beginning and the end of the century, about 20 1999-US dollars, but in between there were some spectacular changes. The price of oil was relatively steady until the first energy crisis in 1973 but on a declining trend; at its low point in 1970 it was slightly above seven dollars per barrel, in 1999-US dollars. Then it quadrupled over a very short time period in the wake of the energy crisis in 1973, and doubled after the revolution in Iran in 1979. It fell quickly, however, particularly in 1986, when Saudi Arabia tried to claw back its share of the oil market. Since 1986 the price of oil has fluctuated with a declining trend in real terms, but at the dawn of the new millennium we may be seeing a reversal of that trend.

Even if the oil price rose beyond 30 dollars per barrel in 2000 it was still only half of what it briefly was, in real terms, in 1980 after the Iranian revolution and way below what many back in the late 1970s believed that it would be in 2000. Partly under the intellectual influence of the Hotelling theory and partly because of the oil crisis in 1973 many pundits believed in the late 1970s that the oil price would rise relentlessly in real terms. The following is an excerpt from a British Green Paper on energy policy from 1978, when the price of oil was slightly above 30 dollars per barrel in 1999-US dollars.[26] "The prospect in the world energy market is that oil, the dominant fuel in the economies of most industrialized nations in the post war period, will become increasingly scarce and expensive during the remainder of the century and beyond, that its price will at least double in real terms by the year 2000." In fact, the high oil prices that the leaders of the European Union and the United States were complaining about in the late summer of 2000 were at about the same level in real terms as when the British Green Paper was written in 1978.

A recent review of commodity prices by the World Bank generally shows declining prices over the last century.[27] The price of coal at the end of the century was lower than at the beginning, but surged upwards in the 1970s and 80s under the influence of the oil price. The price of gold, curiously perhaps, developed in a fashion not unlike the price of oil but was lower at the end of the century than in the beginning. The price of copper fluctuated around 6500 1999-US dollars per tonne from 1900 to the end of the First World War but has averaged about 2500 dollars per tonne over the last quarter of the century. The price of tin fluctuated without much real trend around 15 1999-US dollars per kg until the early 1970s, then surged, and fell to about six to seven dollars in the 1990s, after the collapse of the tin cartel.

The fact that the market prices of minerals have not risen in real terms does not really refute Hotelling's theory, however. As already stated, it is the rise in the rent that counts; this is what the owner of a mineral deposit would invest in some productive asset if he dug his mineral out of the ground, and so it is the rent that would have to rise at a rate equal to the rate of return on productive assets. The rent could rise over time despite a steady or a falling price, provided the unit cost of extraction falls over time. The unit cost of production has undoubtedly fallen, because of technological progress, but reliable data on unit costs are hard to come by, and they differ in any case from deposit to deposit while the prices of the mineral produced from different deposits are much more similar. Since it is very difficult to calculate the rent per unit, among other things because it varies from one deposit to another, it is correspondingly difficult to refute the theory. Still,

---

[26] *Energy Policy: A Consultative Document* (Cmnd. 7101), Her Majesty's Stationery Office (HMSO), quoted in Appendix II, p. 20, in *"The Challenge of North Sea Oil"*, Cmnd. 7143, HMSO, London 1978.
[27] "Commodities in the 20th Century". *Global Commodity Markets*, January 2000, pp. 8-29. The discussion of prices in this paragraph is based on this publication.

this writer would not bet on rescuing Hotelling's theory by investigating how costs have changed over time.

Why have mineral prices fallen in real terms? One reason is the fall in the unit cost of production. Another reason is the substitution of new and better or cheaper materials for old and expensive ones. Oil has been substituted for coal, natural gas for oil and coal, and aluminum for tin and copper. Substitution of renewable resources for non-renewable ones is what will have to happen if the "grand dynamics of humanity" are to strike a happy note. We have already encountered substitution in the peat fable and the story of coal since Jevons. Peat was replaced by coal and became next to worthless, coal was partly replaced by oil and its price was depressed. Maybe the transition from oil and other minerals to new materials and processes is already in its infancy. If that transition takes off, the price of oil and minerals would not rise but fall, and the value of the remaining deposits of these resources would be to some extent or even entirely transient. This makes mineral wealth even more transient than otherwise; not only would it be important to transform it into renewable wealth once it is out of the ground; it would also be important to get it out of the ground while it still is worth something.

Who is entitled to mineral rents?

Who is entitled to the rents arising from the exploitation of natural resources? This is a question of who owns resources, and who is entitled eventually to tax resource rents. There exist different kinds of regimes with respect to ownership of resource deposits. More often than not, the ownership of such deposits is vested in the state itself. In federal states this may be the province, as the case is in Canada and Australia. In the United States mineral rights are private and follow the ownership of the land above, but can be separated from the land and sold or leased. Oil and other minerals underneath state or federal lands are, of course, state or federal property, and all offshore oil in the United States is federal property, as is also the case in Canada and Australia. Irrespective of ownership, governments usually reserve for themselves the right to levy taxes on the extraction of minerals. Again, in federal states they usually share some of this revenue with the province or state in question. Sometimes it is the province or the state which is empowered to tax the extraction of the mineral.

Exactly which individuals are entitled to benefits from mineral rents thus depends on the boundaries of states and provinces, the ownership rights within sovereign states, the division of power between different levels of government, the redistribution of incomes within states, apart from the distribution of minerals across the globe. As to the distribution of minerals this is most uneven. Mineral deposits are concentrated in certain areas where the forces of nature have permitted their formation. The concentration of mineral deposits is perhaps most dramatic with respect to oil and gas. Over 60 percent of all proven reserves of oil are concentrated in just five countries in the Middle East, which jointly account for less than two percent of the world's population.[28]

Sometimes, state boundaries have been drawn narrowly around clusters of mineral deposits. In such cases the resource deposits per capita can be extremely large and the inhabitants of the respective countries correspondingly wealthy, provided the resource deposits are valuable enough. Again oil is the prime example. Kuwait, Brunei, Qatar and the United Arab Emirates are examples of such resource enclaves. Between them they hold almost 20 percent of the world's oil reserves but account for less than 0.1 percent of world population. These are the four largest oil producing countries in the world in terms of production per capita, and their gross domestic product is correspondingly high. The GDP per capita of Kuwait, Brunei and the United Arab Emirates was about 17,000 US dollars in 1998 and depressed by low oil prices, but they still came ahead of Spain, with about 14,000 US dollars per capita. The GDP per capita in Qatar was less, about 10,000 US dollars, comparable with that of Portugal.[29] The contrast with their neighbors is considerable, not to say stark; the GDP per capita in 1998 was about 6,000 US dollars in Saudi Arabia, 3,000 in Malaysia, less than 500 in Indonesia and paltry 260 in Yemen. The populations of Kuwait, Brunei, Qatar and the United Arab Emirates are so small that they have to rely on guestworkers to a large extent. These guestworkers are seldom if ever given rights on par with the citizens of the host country who enjoy the perks from the oil riches in various ways, such as guaranteed employment in the public sector at a premium salary and various other entitlements. This is also true of some other oil rich countries in the Middle East and elsewhere.

Why boundaries have been drawn in this way depends on a complex of factors; tradition, wars, interference of now faded colonial powers, and negotiations among states. In some cases the way boundaries have been drawn is hardly independent of how rich a country is or could be expected to be. The sultanate of Brunei was for a time a part of the British Empire, like its neighboring Malaysian provinces of Northern Borneo, but when the Federation of Malaysia was formed oil had long since become a valuable resource, and the sultan figured that he could do better on his own.[30] Kuwait,

---

[28] These countries are Iran, Iraq, Saudi Arabia, Kuwait and United Arab Emirates. Source on reserves: *BP-Amoco Statistical Review of World Energy*, 1998.

[29] Figures calculated from the World Bank Database for 1998.

[30] See, e.g., Andaya and Andaya (1982).

Qatar, and the United Arab Emirates were all a part of the Ottoman Empire, which disintegrated after the First World War. The European colonial powers, Britain and France, carved up the previous provinces of the Ottoman Empire but with a certain degree of cooperation with the local chiefs, such as the emirs of Kuwait, Qatar and what later became the United Arab Emirates. This was before oil became the major source of energy in the world, but expectations with regard to oil deposits may nevertheless have played a role in how the spoils of the Ottoman Empire were divided. In our time some and possibly all of these owe their independence to the fact that the rule of law has become rather stronger than it used to be in governing relations between states. In earlier times states used to take whatever territories they could occupy and defend but nowadays predatory states that do not respect state borders, or ignore the rights of minorities within their own borders, risk being ostracized or perhaps punished militarily. The repelling of the Iraqi occupation of Kuwait is a case in point, although it would be naive to expect the United States and their allies to have come to the rescue of Kuwait for altruistic reasons.

Interests similar to those that have drawn the borders of sovereign states narrowly, in order to avoid having to share resource wealth widely, also try to bolster the power of resource rich regions within large sovereign states. As already mentioned, the states of Australia and the provinces of Canada hold the rights to their mineral resources, and some American states, most notably Alaska, own oil and other mineral resources. In the United States and Australia this goes back to the framing of the constitution and appears to be uncontroversial. In Canada the western provinces were given the power over their natural resources in 1930, but there has for long been a conflict between the federal government of Canada and the westernmost provinces over natural resources, particularly with Alberta over pricing and export of oil and gas. In non-federal states with little or no regional autonomy it is less easy to restrict the benefits from resource extraction to the inhabitants of a certain region. This sometimes causes conflicts, particularly when the resource rich regions are populated by ethnic groups different from those in other regions or the rest of the country. The Scottish National Party was bolstered by the discovery of North Sea oil, much of which would have been on a Scottish continental shelf if Scotland had been a sovereign state. The island of Bougainville has been at loggerheads with the rest of Papua New Guinea over mining on the island. The oil rich regions of Nigeria, which is a federal state, have a long-standing conflict with the central government of the country.

If one looks for some universal principle determining who is and who is not entitled to rents from resource extraction one looks in vain. There is no such principle; it depends on the development of history, on how each country has chosen to govern itself, etc. With regard to the way the world should look, as opposed to the way it does look, would searching for such a principle be fruitful? From one perspective mineral rents might be said to belong to those who have the knowledge, the foresight, the audacity, and all else needed to get the mineral out of the ground and turn it into a saleable product. This would point squarely at the mining companies. From another perspective these rents might be regarded as our common heritage, a gift of nature to those who inhabit the globe, with minerals having been made valuable by our collective knowledge and development from savages to modern man. But would we not then all be entitled to a share in mineral rents? Why should an Alaskan living in Anchorage or in Juneau be entitled to oil pumped from the uninhabited north thousands of miles away, having done nothing to get it out of the ground? Why should not a Palestinian Arab be entitled to the oil riches of Kuwait just as well as his fellow Arabs who happen to be living on top of the oil fields a few hundred kilometers away? And the distance of a few hundred kilometers is not really an issue here, some Palestinian Arabs have stayed and worked for years in Kuwait, without being given the rights of Kuwaiti citizenship and the perks that go with it.

That notwithstanding, one should not overlook the functional role of a clearly defined and enforced resource ownership. To an idealist, the natural resources, be they non-renewable or otherwise, may look like the common heritage of mankind, to the benefits of which we are all entitled. The troubles start once we try to put such lofty ideals into practice. Somebody has to have the incentive to get oil or whatever it is out of the ground and to market. That, incidentally, is not just a matter for an oil company to do; there must be some infrastructure in place, and a security for the oil company that its rights of ownership or leasehold will be honored, which in turn presupposes a civil society of some degree. That in itself may be a functional argument for taxing away some of its rents and to redistribute them, to gain acceptance of the activities of the mineral extraction companies in the societies in which they are embedded.

CHAPTER THREE
MINERAL RENT INVESTMENT FUNDS

"The purpose of income calculations in practical
affairs is to give people an indication of the
amount which they can consume without
impoverishing themselves. ... (I)t would seem
that we ought to define a man's income as the
maximum value which he can consume during a
week, and still expect to be as well off at the end
of the week as he was in the beginning."

Sir John Hicks (1946): *Value and Capital,* 2nd
Ed., p. 172.

As the above quotation shows, the British economist and Nobel
laureate Sir John Hicks defined income as the maximum value a man can
consume during a week and still expect to be as well off at the end of the
week as he was in the beginning. Money withdrawn from one's bank
account in excess of the interest it yields is not income, and financing one's
consumption out of one's savings is not a sustainable activity, as most people
know well enough. These elementary ideas have a straightforward
application in mineral economics. Given that mineral resources are finite,
exploitation of such resources is not a sustainable activity. By investing an
appropriate share of mineral rents in productive assets it is possible to
preserve mineral wealth, albeit in a different form, and to turn the
exploitation of mineral resources into a sustainable activity in the sense of
permanently raising the standard of living.

Consider the following very simple example where a portion of
mineral rents is invested in financial assets with a fixed rate of return (a bank
account or bonds). Suppose we have a mining activity capable of yielding a
rent of 100 million dollars per year for ten years. We ignore the possibility
that the rent might rise over time, which was discussed in the previous
chapter, so there is nothing to be gained by postponing the extraction. By
investing a certain share of the rent we could build up an investment fund
which would allow us to spend the amount X per year in perpetuity, even

long after the mining activity has come to a halt, provided we maintain the fund. How much would we have to save? We can find this in a simple way. Each year we would spend X and put aside $100 - X$ in an investment fund. If the rate of return on the investment fund is $r \times 100$ percent, the fund would have grown to

$$F = [100 - X](1 + r)^9 + [100 - X](1 + r)^8 + ... + 100 - X$$
$$= \frac{[100 - X]\{(1 + r)^{10} - 1\}}{r}$$

at the end of the ten year period. From then on we could use $r \times 100$ percent of the fund annually without dipping into the fund. This is X, the amount that could be consumed in perpetuity. Setting X equal to rF gives

$$X = 100\left[1 - \frac{1}{(1 + r)^{10}}\right]$$

Assuming a five percent rate of return ($r = 0.05$), we find $X = 38.61$. In other words, saving every year just over sixty percent of the mineral rent would, in this particular example, turn the mineral wealth into a permanent wealth that would make it possible to spend a value equal to just under forty percent of the annual mineral rent in perpetuity, even if the mineral rents would stop flowing after ten years. Reality is, of course, a lot more complicated than this simple example; the flow of mineral rents is not even, it is not known exactly ahead of time, and the rate of return is also variable and uncertain. Furthermore, we need not opt for a constant consumption per year in perpetuity; it might be more desirable to go for an increasing consumption over time, for example, if the population is expected to grow. In that case we would have to constantly invest some of the return on the fund (rF) to keep it increasing. We will meet an example like that in the following chapter on Nauru. The principle, however, remains much the same despite these complications, it is just more difficult to put it into practice, and in the real world where there is uncertainty about the future one may not exactly achieve what one set out to do.

We can make this example slightly more general by assuming that the extraction period lasts for T years. Our formula would in that case be

$$X = 100\left[1 - \frac{1}{(1 + r)^T}\right]$$

We can get a certain feeling for the importance of the two parameters, the interest rate and the length of the extraction period, by looking at the X-value produced by alternative constellations of parameters. A sample is shown in Table 3.1 below. If the rate of interest is as low as one percent and the extraction period is only ten years, more than ninety percent of the annual rent must be saved. The effect of changing the rate of interest is dramatic. With a ten percent rate of interest, less than forty percent of the annual rent need be saved to provide for a sustainable consumption amounting to more than sixty percent of the annual rent. And remember, the rents are collected for only ten years while the consumption goes on for ever.

Table 3.1. Optimal consumption, in percent, from an even flow of rents for T years, at $r \times 100$ percent rate of interest

$r \downarrow$ $T \rightarrow$	10	50	100
0.01	9.5	39.2	63.0
0.05	38.6	91.3	99.2
0.1	61.4	99.1	100.0

With rents that are being collected at an even rate for 100 years the savings become significantly smaller. With a rate of interest of only one percent it would be necessary to save 37 percent of the annual rent to produce an even, sustainable consumption. And with a ten percent rate of interest it looks as if nothing needs to be saved, but this is a rounding off effect; the correct solution with four decimals is 99.9927, which gives a savings rate of 0.0073 percent. But the power of compound interest is strong; savings of only 0.0073 in year one will have grown to 100 after 100 years at a ten percent rate of interest if left untouched.

Needless to say, the real world is in many ways different from this simple example, even if it suffices to illustrate elementary principles. The even stream of rents over a given time period can be thought of as the product of a given volume extracted per year and a constant price net of extraction costs. Mineral income streams of the real world, apart from being uncertain and subject to fluctuating prices and changes in technology, are likely to be different. In the beginning the income will probably be low or negative while money is being invested in the production equipment, but after the investment phase has been completed the income stream will reach a peak and possibly taper off after that. This is typical, for example, of petroleum production. In all probability it would be deemed desirable to smooth over time the consumption to be financed by this income stream. Initially debt would be accumulated, but later it would be paid off as the income rises and an investment fund gradually built up, the income of which would replace the mineral income once the mineral runs out, much as in the previous example.

There is a simple rule of optimal saving of mineral rents in case we want to achieve sustainable consumption. Let the present value of the mineral wealth be denoted by W, and the sustained annual consumption by X, as before. The present value of an annual consumption of X, sustained in perpetuity, is X/r, where r is the rate of interest.[31] This present value must be equal to the present value of the mineral wealth; that is, $X/r = W$, which implies $X = rW$. Hence we can consume five percent of the present value of the mineral wealth if the rate of interest is five percent, etc., which means that we must save 95 percent of it.[32] The optimal savings plan would be to save most when the rents are high and borrow when the rents are low, an implication of the desire to smooth consumption over time. If the time profile of rents is such that most of them come early, as is typical of oil production, for example, most of the savings for sustainable consumption would have to be undertaken early.

The rule that tells us that we can consume annually an amount equal to the rate of return on our wealth has been called the Hicksian Rule, after the British economist Sir John Hicks, with whose definition of income we started this chapter. Applied to mineral wealth, the income from extraction of minerals is equal to what we can consume out of our mineral revenues after having invested enough to make sure the mineral wealth is preserved.

One possible savings plan that would build up a fund equal to the present value of the mineral wealth is to save a certain fraction of declining mineral rents, provided the savings will not be touched until the mineral rents have run out. Note, however, that if the return on the fund would be used right away, only a fraction of the mineral wealth would be saved; the reason

---

[31] This comes from the formula for infinite geometric series. The present value of consuming X per year in perpetuity is $X/(1+r) + X/(1+r)^2 + \ldots$, which can be shown to have the finite value X/r even if the series is infinite.

[32] If the reader finds these savings high, compared to the savings implied by the figure in the last column of Table 3.1, he or she is again tricked by the logic of compound interest. The present value of all rents is $100(1+r)^{-1} + 100(1+r)^{-2} + \ldots + 100(1+r)^{-100} = 100[1-(1+r)^{-100}]/r$, which comes to 1984.79 with $r = 0.05$. The accumulated value of saving S each year for a hundred years is $S(1+r)^{99} + S(1+r)^{98} + \ldots + S = S[(1+r)^{100}-1]/r$, which comes to 1984.79 with $S = 0.7605$. And, five per cent of 1984.79 is 99.23955, the sustainable annual consumption, as shown (rounded off) in the table. The remaining 95 percent of the present value of the rents is saved and invested. The low annual savings come about because the savings process goes on for a hundred years, and the incomes from the investment fund built up with these savings do not have to be touched until after hundred years but can be added to the fund.

why it is enough to save only a fraction of the mineral rents and still preserve the entire mineral wealth is the leverage provided by investing the return on the savings already made while the mineral is being extracted. The state of Alaska has decided to save a certain fraction of its oil revenues but not all the return on the fund is reinvested. The province of Alberta likewise opted for saving a certain fraction of its oil and gas revenues in the 1970s and invested the return on the savings, but this plan has long since been abandoned. We return to the cases of Alaska and Alberta in Chapters Five and Six, respectively.

How much of a declining rent would have to be saved in order to preserve mineral wealth? Suppose the flow of mineral rents declines geometrically at the rate k, so that the rent in period t (R_t) is equal to

$$R_t = \frac{R_0}{(1+k)^t}$$

Saving the fraction s for T years would build up a fund (F) that, at the end of T periods, would amount to[33]

$$
\begin{aligned}
F &= sR_0 \left[(1+r)^{T-1} + \frac{(1+r)^{T-2}}{1+k} + ... + \frac{1}{(1+k)^{T-1}} \right] \\
&= \frac{sR_0(1+r)^{-1}\left[(1+r)^T - (1+k)^{-T}\right]}{1-(1+r)^{-1}(1+k)^{-1}}
\end{aligned}
$$

The present value of the flow of rents (the mineral wealth, W) is

$$
\begin{aligned}
W &= R_0 \left[(1+r)^{-1} + (1+r)^{-2}(1+k)^{-1} + ... + (1+r)^{-T}(1+k)^{-(T-1)} \right] \\
&= \frac{R_0(1+r)^{-1}\left[1 - (1+r)^{-T}(1+k)^{-T}\right]}{1-(1+r)^{-1}(1+k)^{-1}}
\end{aligned}
$$

Setting $F = W$ gives

$$s = (1+r)^{-T}$$

---

[33] The rent each period is assumed to accrue at the end of each period, and so the amount invested in the first period will have accumulated interest over T-1 periods at the end of the extraction.

The fraction to be saved is inversely related to the rate of interest and the length of the extraction period. This is easily explained; the higher the interest rate the higher the return on the savings put into the investment fund, and the lesser the need for saving out of current rents. And the longer the extraction period, the longer will the accumulated savings remain in the fund before being used to finance current consumption, and the less the need to save from current rents. In this case the necessary savings rate is independent of the decline rate of the rents, but this is an artifact of the geometric rate of decline.

Table 3.2 gives an idea of the sensitivity of the savings rate to the rate of interest and the time of extraction. With an interest rate of only one percent it would be necessary to save ninety percent of the rents if the extraction period is only ten years, but the savings rate drops to sixty percent if the extraction period lasts for fifty years. With a ten percent rate of interest it would only be necessary to save about forty percent of the rents if the extraction period is ten years, and tiny one percent if it is as long as fifty years. Again, note that this requires that all the return on the fund will be reinvested in the fund until the extraction period is over. If, instead of a financial fund, we think in terms of investing in machines or other real assets, this would amount to saying that all the return on those assets should be reinvested until the mineral extraction is over.

Table 3.2. The necessary constant savings rate for preserving mineral wealth with geometrically declining rents.

T↓ r→	0.01	0.05	0.1
10	0.905	0.614	0.386
20	0.820	0.377	0.149
50	0.608	0.087	0.009

A constant savings rate is unlikely to produce an optimal consumption profile if mineral rents decline over time. This would result in a profile where consumption is largest initially, declines gradually and reaches a minimum at the end of the extraction period, and then jumps up to the sustainable level as the consumption becomes equal to the return on the fund built up from the mineral rents. Figure 3.1 compares this consumption profile with a sustainable consumption of mineral rents, according to the rW rule. According to this rule, the amount rW is consumed each year while the rest is invested. This would give a constant consumption during the extraction phase, while the investment of the remaining part of the rents, and reinvestment of the return on the savings, would suffice to build up a fund equal to W at the end of the extraction period.

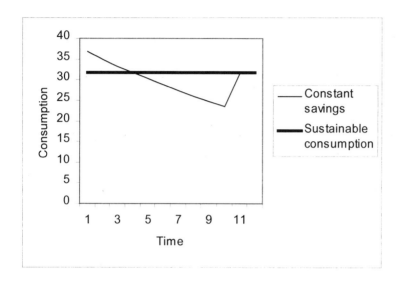

Figure 3.1. Consumption with a constant savings rate for mineral rents versus sustainable consumption from mineral wealth. $k = 0.04$, $r = 0.05$, $T = 10$, $R_0 = 100$.

Saving rents: by whom and for whom?

Establishing the principle of optimal saving of mineral rents is one thing, but how this could be accomplished is a different question. Who is best to be trusted to save rents optimally? The mineral companies? The government? Individual households, after rents have been taxed away and distributed among the citizenry at large? Perhaps this would be best left to the mineral companies themselves. Companies are in business to make money for their owners, they seek new and profitable opportunities to do so better still; indeed their managers risk being fired or unfriendly raids on the company stock if they do not. Hence, if the mineral rents remain in the mining companies they would probably be invested in new and profitable ventures. One cannot exclude, however, that at least some of the rents would be used for featherbedding the management and even ordinary workers and salarymen. The stock exchanges do not necessarily possess complete information on how well a company is run and what might be the prize of a successful raid on its stock.

But the question who should invest mineral rents can hardly be answered without also asking the question for whose benefit? As already mentioned, the rents from some mineral deposits, oil in particular, can be spectacularly high. It is not acceptable, at any rate in democratic and egalitarian societies, to have private companies collect all this profit for the benefit of their shareholders. This is, of course, particularly true in countries where the mining company is in the hands of foreign shareholders. The company might well invest all the rents with the utmost care for its shareholders, but little if anything might trickle down to the citizens of the country where the mining operation is located.

To make sure the rents are invested for the benefit of a broader constituency than the shareholders of mining companies, the government of the nation or the province entitled to the resource has to step in and collect the appropriate share of the rent. This raises two questions. Can the rent be collected without harming the mining operation? Taxing mining operations to death would amount to killing the goose that lays the golden egg. Second, can governments be entrusted with investing the appropriate share of mineral rents?

In principle it is not too difficult to cream off a substantial share of mineral rents by taxation. The rent is a residual; it is whatever is left when all costs have been paid. This makes mineral rent an almost perfect object of taxation. In principle it would be possible to tax away practically all the rent; the mining firm would still be interested in extracting the mineral as long as it covers all its costs, including capital cost. In practice taxes on mineral companies are seldom designed in this way. Special mineral taxes are sometimes levied on gross income, providing incentives for companies to cut off production too early when the unit cost rises as reserves are depleted, as is typically the case. In other cases they are levied on net income defined in a way that does not appropriately account for all costs of production, as usually is true as well of the ordinary corporate tax; the costs may be exaggerated through over-generous amortization rules or they may be insufficiently accounted for, e.g., by excluding financial costs from deductible costs. This distorts the incentives faced by firms; they will either invest too little or pay insufficient attention to saving costs and invest too much.

The closest the real world has ever come to a tax on mineral rents is the so called resource rent tax practiced for some oil fields in Australia and which was, in fact, invented by two Australian economists, who apparently managed to become prophets in their own country.[34] One possible reason why resource rent taxes have not caught on despite their theoretical advantages are the formidable problems in correctly estimating the rent to be

---

[34] See Garnaut and Ross (1975, 1983).

taxed. The tax returns of mining companies can be "doctored" in various ways. The companies may be buying goods and services from or selling their product to their own subsidiaries and could therefore set prices as appropriate for reporting their profits in jurisdictions where the tax regime is lenient. Another problem is the capital cost, which consists in part of the rate of return the companies could earn if investing elsewhere, a quantity that is likely to be difficult to assess.

As to how governments will spend the income from taxing mineral companies, it is not a foregone conclusion that it will be invested as appropriate for future benefits. Governments are under a certain pressure to spend for immediate benefits rather than save for the more distant future. In part this comes from the short-sightedness of the electorate; future generations are not yet around to vote for those who best serve their interests, and it is quite likely, therefore, that the popularity of elected politicians will be better served by spending for the short term rather than saving for the long term. But politicians and empire-building civil servants also may have motives that would be better served by spending now rather than saving for a distant future, or by investing in prestigious projects that have a low return. Landing a research institution, a bridge or a highway in one's constituency is likely to enhance the chances of reelection for the politician who brings these goodies. Some research facilities and institutions in the United States have gratefully been named after the senator who made sure the money was available. Donations to foreign lands or institutions will increase the standing of the visiting politician or head of state who brings them along. Last but not least, politicians and officials in countries with corrupt governments with little or no democratic control use government money to buy influence and line their own pockets. Much of the mineral wealth of Nigeria, Congo (for a while known as Zaire) and Russia has not been saved for the benefit of their peoples but for a corrupt elite. Entrusting the savings of mineral rents to governments has its own risks, both in democratic and undemocratic societies.

The imperfections of the world and its governments are things we must live with. To mitigate them, it is imperative that a system of incentives be set up such that it will be in the interest of governments to take the appropriate decisions, in this case save the appropriate share of mineral rents for the future. This would probably be best served by channeling all income from taxes on mineral extraction into an investment fund. This would make the income from mineral extraction visible and separate from other government income, which is a first and necessary step to ensure that it would not simply be spent on current consumption and then be gone. Government income from mineral extraction is by its nature quite different from other government income. The ordinary business of governments is to provide public services of various sorts, and for this purpose they need income, to be financed from the ongoing activity in the economy. Mineral rents are non-renewable wealth which needs to be transformed into renewable wealth before it yields any income to be spent. The case for having governments collect mineral rents through taxation is to make sure they are invested for a wider constituency than the shareholders of mining companies, not to spend them on unsustainable public consumption.

The most effective way to give politicians incentives to build and maintain a mineral rents fund and to maximize its long term return is probably to give the electorate a direct stake in the fund. The electorate would then put pressure on their representatives to preserve the fund and see to it that it is managed properly. There are various ways in which the electorate can be given a direct stake in the fund. One is to distribute its return among the electorate at large, as is done in Alaska; the real income of the Permanent Fund is distributed among all Alaskans. This seems to have entrenched the fund firmly with the Alaskan electorate. Another way would be to use a mineral rents fund as a retirement fund. These issues will be further discussed in later chapters.

In what should the rents be invested?

Various rules and regulations are needed to make mineral rents investment funds function properly. Elementary questions are how much and in what to invest. As demonstrated in the simple example above, the appropriate share of mineral rents to be invested depends on the time profile of the mineral rent income and the rate of return on investment. In an uncertain world where both the expected time profile of income and the rate of return on investment, and other relevant factors, change over time in unpredictable ways, rules on how much to invest should ideally be contingent on different states of the world. Such practical examples as there exist of investment funds financed from mineral rents are less sophisticated, however, in this respect. Both in Alaska and Alberta a certain proportion of mineral tax income has been channeled into an investment fund, rather than being made contingent on specified events. In Alaska further extraordinary allocations have been made from time to time. In Norway all government income from oil and gas extraction goes initially into the Petroleum Fund, but how much in fact is invested depends on how the government budget is balanced.

Having decided how much to invest, rules will be needed for what to invest in. The most basic choice is probably whether the fund should be invested in assets that yield a visible pecuniary return or in less tangible wealth that may not be measurable or yield a pecuniary return. This is by no means a straightforward choice. The productive capacity of an economy does not depend solely on investments in buildings and machines; it depends no less on the educational level and the health of the labor force, and infrastructure such as roads, airports and telecommunications. Investment in education, health and infrastructure is seldom left to the private sector and, when that happens, only partially, because the private return on such investment usually is much too narrow a criterion for success; such investment is typically the responsibility of governments. Income from taxing mineral extraction will make it possible to undertake more such investment.

Worthy as investment in infrastructure, health and education may be, there is reason for being cautious in using mineral rents for these purposes. Precisely the fact that the narrow but clear criterion of a pecuniary return on investment cannot be relied upon to judge the desirability of such spending opens the floodgates for abuse. If mineral rents are channeled into a fund to be used for these purposes one can be sure that there will be no end to worthy projects being peddled by individuals or groups with an interest in having them carried through, irrespective of their merits otherwise. Precisely because the benefits of such spending are to a degree hypothetical there will be a certain leeway to interpret them in a way that suits one's interests. Consultants willing to provide the necessary figures and other window dressing are not in short supply. There is a real danger that if mineral rent investment funds are to be spent for such purposes they will end up making a very limited contribution to the goal of economic growth that they were established to promote. The track record of the Alberta Heritage Fund, to be discussed below, lends some support to that statement.

It may be argued that this criticism hits all government spending on education, health and infrastructure and that, if taken to its logical conclusion, one should argue for getting government out of such undertakings altogether. But the market has its own shortcomings in this area as well; certain things cannot be left to markets to solve, even if it is debatable which and to what degree. Ideally, the level of public services should be determined by equality between their marginal benefit and their marginal cost, but the problem is that there is no mechanism, as there is for private goods sold in markets, ensuring that those who make decisions about the level of public services are the ones who at the same time get the benefit and pay the cost. When public services are being paid for by taxes the reluctance of the public to pay taxes serves as a proxy, albeit an imperfect one, for the marginal cost of those services which, one may surmise, politicians might just as soon provide until there is no benefit at all, because they do not pay the cost but gain politically from having provided services which at least some of their constituents find useful.

The number of tunnels, bridges and prestigious airports that could be financed out of mineral rents to bolster the reelection chances of politicians is potentially large. What makes mineral funds special is that using them for such purposes would be an almost painless undertaking, for politicians and the public alike.

Hence, it is probably wise, and supported by the case studies to be reported below, to give mineral rent investment funds a mandate to invest their money in projects that are profitable on the basis of conventional market criteria. There are, needless to say, many technicalities that still would need to be resolved. A higher potential yield typically requires taking greater risks. What would be the optimal balance between risky and less risky assets? These are technical questions which will not be dealt with here. A mineral rent investment fund should probably be organized as an autonomous institution, at arm's length from the ordinary business of government and independent of the legislature, except insofar as all institutions in a democratic society are ultimately under the control of the legislature. The Alaska Permanent Fund is an example of such an institution. The managers of an investment fund with a mandate to invest in profitable projects, as judged by realized monetary rates of return, could be evaluated on the basis of objective criteria and kept to their mandate by contracts specifying the terms of hiring and firing. Both the Alaska Permanent Fund and the Norwegian Petroleum Fund employ professional broker houses which must satisfy certain objective criteria with respect to rate of return and how it varies over time. Funds not subject to such criteria and hidden away from the informed public would be liable to mismanagement or corruption. One such case is the Nauru phosphate funds, of which more in the next chapter.

Arguments may, however, be advanced against the need for and even the instrumentality of investment funds as devices for saving incomes from taxes on mineral extraction. Earmarking a certain share of mineral rents for investment may not be quite what it seems. Oil money is not any different from other money; it does not smell of oil, and it is not greasy at the edges or otherwise stained; in fact it is just entries in some bank account. The point of this is that a government's use of mineral rents has to be seen in the context of public finances in general; little would be won by investing mineral rents if the government runs up a debt on its remaining activities; what matters is the total level of expenditures and the total amount of investment. This was the argument made by some Norwegian officials against the idea of setting up the so-called Petroleum Fund, and we will see its relevance below in the fate of the Alberta Heritage Fund. That notwithstanding, it would seem that the obligation to set aside some of the mineral rents into an investment fund and maintain the real value of that fund would impose a tighter constraint than otherwise on imprudent government spending and the buildup of public debt, particularly if the fund is an autonomous institution independent of the government and inaccessible for paying its debts.

If properly managed, an investment fund financed by mineral rents will yield a return comparable to what otherwise will be obtained by financial institutions in the markets it invests in. The transformation of the mineral wealth to permanent wealth will not have been accomplished until a fund equal to the initial present value of the mineral rents has been built up. If some of the return on the fund is used before that goal has been obtained, a larger share of the rent will have to be invested in each period, but once the fund has reached the level of the initial mineral wealth all of the return can be used without encroaching upon the wealth.

How should the income of a mineral rents investment fund be used, once it has been built up? One possibility would be to help pay the ordinary expenses of government; investing in infrastructure or education, or lowering taxes. The main argument in support of this is that the return on the fund would not impose the burdens on the economy that ordinary taxes do. Taxes on labor income discourage work effort, taxes on capital income discourage investment, and taxes on goods and services reduce the total amount used and provided of the goods or services involved. Financing public expenditure from the income of an investment fund would have none of these effects, it would be a painless source of government income. But herein lies also the downside of this method of financing government expenditure. We have already discussed how there would be a tendency for overprovision of public services if they could be painlessly financed. An alternative way of using the income of the investment fund is to somehow distribute it to the citizens at large and let them spend it for their own direct benefit. One advantage of this approach is that the ordinary citizen would have a direct interest in the buildup and maintenance of the real value of the fund and the maximization of its return in a long term perspective, which would put pressure on elected politicians to this effect, as discussed above.

Where to invest, at home or abroad?

Another fundamental choice with respect to investing mineral rents concerns whether to invest them at home or abroad. The answer to that question hinges on at least three considerations, the size of the domestic economy, the need to diversify away from the mineral sector, and the level of development of the economy. The two first ones are related; the smaller an economy the more likely it is that it will be dominated by the mineral sector. Given that the prices of minerals are quite volatile, the need to diversify will be great if the mineral sector is dominating. In addition, the mineral reserves may not last very long.

It may appear self-evident that an economically underdeveloped country ought to use its mineral wealth to promote its own economic development by investing in profitable ventures inside its own borders. Many large and populous countries in the world are well endowed with natural resources but poor in terms of GDP per capita. Such countries could accelerate their economic development by a judicious use of their mineral rents. The best way of doing so need not be investment in industrial equipment; investment in infrastructure and education might be more conducive for economic development. Some countries have made good use of this opportunity. A case in point is Botswana and its use of its diamond wealth.[35] Sadly, however, many and perhaps most other poor countries do not seem to have succeeded in taking advantage of the opportunities their mineral wealth offers. Often the mineral rents have largely been spent on unsustainable consumption. Sometimes the problem has been the wrong type of investment rather than insufficient investment; Algeria and Venezuela invested some of their windfall gains from oil in the 1970s in unprofitable industries.[36] This is particularly likely to happen if the government itself owns these industries, as government decisions on such matters are less subject to the discipline of the market and the need to make a profit. And then there are the cases where mineral rents have been appropriated by a corrupt elite and salted away in distant lands offering the desired secrecy with respect to bank accounts.[37]

So, in order to use mineral rents to promote economic development there must be in place a mechanism which channels the money towards this end. A reasonably honest and efficient administration is of course essential, but apart from that this is in part a question of the degree of development of financial markets. If financial markets are well developed they will provide information on profitable investment opportunities not just to a mineral rent investment fund but to all investors. But if this is the case there will in fact be little reason for the fund to especially target the domestic economy; other sources could be relied upon for investing in such projects as are in fact profitable. It could even be argued that a mineral rent investment fund ought to refrain from investing in the domestic economy and rather rely on the private sector and foreigners to do so, as this would be the surest way of having profit opportunities in the domestic economy judged by an unbiased

---

[35] See Auty and Mikesell (1998).

[36] See Gelb and associates (1988).

[37] A case in point is Nigeria. Its previous ruler, General Abacha, is believed to have siphoned off more than four billion US dollars during his four year reign. On this, and the money laundering involved, see the Financial Times, September 5 and 6, 2000.

party. The country might also be well advised rather to target foreign investment opportunities in order to spread risk and to avoid overloading the domestic economy. This is the option taken in Norway and Alaska, but the latter can be regarded as a small but extremely open economy with a particularly easy access to the contiguous 48 states.

Small countries that we might call "resource enclaves" are a special case. These are countries whose boundaries have been so narrowly drawn that there are very few opportunities outside the mineral extracting sector. Kuwait and Brunei are examples of such resource enclaves. Another is Nauru, which we will discuss in the next chapter. For such resource enclaves there are few other opportunities than to invest in financial assets abroad, providing them with future income derived from the fruits of other people's work rather than their own efforts. Even more sizable countries or provinces, with populations in the millions or large land masses, or both, such as Alaska and Norway, have found it prudent to travel down this road and to invest some of their mineral wealth abroad, because of limited opportunities at home and a desire to spread risk.

But to what extent is it possible for a state or a province, or even for a sovereign country like Norway, to invest mineral rents for the purpose of enhancing the wealth of its citizens? There are no restrictions on migration within the United States or Canada, and migration is now free between Norway and the member states of the European Union. Since people can be expected to migrate in response to differences in income and wealth between countries, states or provinces, attempts at wealth enhancement in one particular area will be to some extent self-defeating in the sense of expanding the population base with the possible consequence of diluting the wealth in per capita terms, unless the in-migrants bring new wealth or wealth generating power with them. In particular this would be expected to result if the wealth enhancement takes the form of entitlement programs doling out income or benefits without any effort being asked in return. For wealth enhancement in resource enclaves to be successful there may have to be in place some system for restricting in-migration motivated by a desire to get a share of the wealth. Some resource enclave economies, in the Middle East for example, deal with this by allowing import of guestworkers without giving them a share in the privileges enjoyed by formal citizens.

CHAPTER FOUR
NAURU AND SUSTAINABILITY

"And what will happen when the phosphate runs
out?"
"See, *that* I would also like to know,"

From a conversation with a young Nauruan,
June 2000.

The island of Nauru illustrates the issue of sustainability well, and some would say starkly. Nauru is an atoll in the central Pacific, about 21 square km, with a circumference of about 19 km. A normal person can walk around the entire island at a leisurely pace in about three hours. Nauru is also among the smallest sovereign states in the world, being inhabited by about 11,000 people, of whom about 8,000 are citizens of the state of Nauru and the rest are expatriates, mainly from other Pacific islands and China.

Underneath the topsoil of most of the island there is, or was, a layer of soil rich in phosphates. In pre-colonial days this was of little consequence for the Nauruans. Before any white man had set foot on the island the Nauruans lived off "coconut, sunshine and fish" and other gifts of nature such as the fruit of the pandanus tree, which used to be something of a staple in Nauru and many other Pacific islands. In fact the interior of the island where the phosphate deposits are located was little used by the Nauruans; they lived along the coast where they did their fishing and cultivated their palms and pandanus trees.

Nauru was among the last outposts to be colonized by Europeans.[38] It was not considered very important, being remote and thought to yield little in terms of natural resources; in fact it was barely able to sustain its original population. Seafarers would occasionally come ashore for water and recreation, and occasionally a white man would remain rather than return to his homeland or to the boredom and toil of his ship. The island had a rather pleasant image, being called Paradise Island, but European colonial powers did not for a long time bother to annex it.

---

[38] The story of Nauru is vividly told in Weeramantry (1992).

In the 1870s and 80s the Germans started to harbor ambitions to become a colonial power. They were, for a number of reasons, late in the game. Germany did not really emerge as a major power on the European scene until the latter half of the 19th century when Bismarck succeeded in uniting Germany under the leadership of Prussia. Bismarck himself was not much of a colonialist but had to yield to powerful forces which thought that a great European power had to have colonies. But there were not many areas that had not been claimed already by other European powers. In the 1880s the last spoils were divided between the European powers. Germany got some leftovers in Africa (Tanganyika, Southwest Africa) and in the Pacific. One of their Pacific trophies was Northeast New Guinea and surrounding islands, and on maps one can still find the name Bismarck Archipelago being applied to a string of islands scattered northeast of New Guinea. Nauru was one of the places given to the Germans at a conference in Berlin in 1886 on European spheres of influence in the Pacific.

It was not the Germans, however, who developed phosphate mining in Nauru. The story of the discovery of the phosphate deposits on Nauru is a strange one. Somebody observed a stone used as a doorstopper in an office in Melbourne. He thought the stone funny and took it to a chemist for analysis. The stone turned out to be extremely rich in phosphate. The interest in the stone was further aroused. Australia was an agricultural country but its soil was poor in phosphates, so good results in farming depended on applying fertilizer. Phosphates from a reasonably nearby source would easily find a market in Australia.

But where did the stone come from? The available information pointed to Nauru. Was there a lot of rock like that there? A small group of prospectors set out to investigate. But Nauru was a German territory, and it was important not to let the Germans find out that they were sitting on an extremely valuable resource. As the prospectors approached Nauru an engine trouble was faked, and they went ashore ostensibly to pass away the time while the engine was being fixed but using the opportunity to look for phosphate deposits. Their expectations were fulfilled.

Negotiations with the Germans about mining rights were successful, and in 1906 the Pacific Phosphate Company, mostly British owned but also involving some German capital, began mining the phosphate. In those days the mining was done with simple hand shovels; the topsoil was removed and the phosphate hacked and shoveled away. For this purpose foreign labor, mainly from China, was imported and kept confined away from the islanders and their European masters.

When the First World War broke out the Australians occupied Nauru. The fate of the island was settled in the peace negotiations at Versailles. Not surprisingly perhaps, the victors of the war found Germany unworthy of having colonies. Rather than taking these colonies over outright they were awarded in trusteeship under the League of Nations, the unsuccessful forerunner to the United Nations. A great row developed at Versailles between the Australian prime minister Hughes and the American president Wilson, whose ideas about self determination did not cut much ice with the colonial powers at the time, about who should get the trusteeship over Nauru and on what terms. The Australians were particularly interested in Nauru because of the phosphates. In the end Great Britain, Australia and New Zealand were given a joint trusteeship over Nauru, with Australia being in charge of the place. The three governments bought the rights to mine the phosphate from the previous holder, the Pacific Phosphate Company, and vested these rights in a body that came to be known as the British Phosphate Commission.

The League of Nations trusteeship passed to the United Nations after World War Two. This put some, if lenient, constraints on the use of executive power in the territory. Missions were sent to the island to investigate how the trusteeship was being carried out, and occasionally it would be debated at the United Nations. These episodes abound in ironies. The Soviet representatives at the UN Trusteeship Council would ask rhetorical questions about the self-determination of the Nauruans and deplore the destruction of the island by the colonial powers. None of them seem, however, to have cited the track record of the Soviet Union with respect to its ethnic minorities and their homelands as examples to be followed. The colonial powers played their own games, however. The Trusteeship Council wanted to know if the Nauruans were getting a reasonable share of the phosphate riches. For this it was important to know the phosphate rent, i.e., the difference between the market price and the cost of excavation. The British Phosphate Commission found it extremely difficult to provide any such information. There was no such thing as a world market price for phosphate, it was said; phosphate was mined in different parts of the world and sold to different markets, and its quality varied from place to place. As to costs, the ones attributable to Nauru could not possibly be distinguished from the commission's operations on another phosphate island, Banaba or Ocean Island. It later turned out that the commission kept detailed accounts of its operations in various places, for its own purposes.

In the colonial period, including the trusteeship, the phosphate mining activities were of little importance for the Nauruans. They got a small royalty from the phosphate mining but were otherwise of little consequence for the mining operations and probably just in the way. There was at one point talk of resettling them somewhere else; an island off the Queensland coast was inspected and offered for this purpose. Something similar did in fact happen to the inhabitants of Banaba, alias Ocean Island, which also was mined for phosphates. The Banabans were removed and resettled on an island in the Fiji archipelago which was bought with phosphate money. The resettlement of the Banabans may have been easier to accomplish because Banaba was a part of what used to be called Gilbert Islands, a British colony, while Nauru was held in trust and overseen by the United Nations.

In 1968 the colonial period came to an end, Nauru got full independence, and the Nauruan government took over the phosphate mining. For this purpose the young state set up its own mining company, the Nauru Phosphate Corporation, which took over the facilities of the British Phosphate Commission. The end of the phosphate mining was then foreseen; the remaining deposits were believed to last for one or two decades. The phosphate mining is, however, still going on but its output rate has fallen precipitously in recent years, from an average of 1.58 million tonnes per year in the 1980s to 0.51 million tonnes in the 1990s up to 1997.[39] The reserves left are certainly limited; the excavation going on at the turn of the millennium was eating up marginal areas such as strips previously left for roads in the phosphate bearing area.

After the Nauruans took the phosphate mining into their own hands they became the beneficiaries of this operation. What had long been suspected turned out to be true, the phosphate mining was extremely profitable. This made it possible for the Nauruans to enjoy a standard of living quite atypical of the Pacific islands. In the late 1980s the GDP per capita was as high as 9,000 Australian dollars, equivalent to 7,000 US dollars or more.[40] Earlier it appears to have been higher still; one source from the early 1980s mentions 20,000 US dollars.[41] For comparison the GDP per capita in neighboring Kiribati, which has no phosphate but more people (85,000), is about 650 US dollars.[42] An American journalist visiting Nauru in the early 1980s described it in the following way:[43]

"Imagine an island nation where the typical family of five has an income of $100,000 a year without even working; where there are no taxes or duties; where all medical and dental care is free, and even medicine, bandaids and aspirin are dispensed without charge."

[39] Asian Development Bank (1999).
[40] Asian Development Bank (1998). All values in this chapter are in Australian dollars, the currency used in Nauru, unless otherwise stated.
[41] Rod Nordland: "An island paradise where one's every need is met", Philadelphia Inquirer, p. 2A, Dec. 13, 1981.
[42] Calculated from the World Bank Database for 1998.
[43] Rod Nordland, op. cit.

The phosphate rentier economy

The phosphate riches have been distributed among the Nauruans in various ways. Virtually all Nauruans have been employed in the public sector, either by the government directly or by the government owned Nauru Phosphate Corporation. In 1998 there were 3561 people employed in Nauru. Only 150 were employed by the private sector, the rest were employed by the government or government corporations.[44] There is little doubt that some of this employment was for the purpose of income distribution rather than need for the work being done. The government has provided every family with a house, and health service and education has been free. The free health service did not just cover the limited services available on the island but also dispatch overseas of cases which could not be treated locally. In short, the phosphate financed a welfare state, and the phosphate wealth seems to have been shared among all Nauruans, although some have undoubtedly got more of it than others, such as owners of land where phosphate has been mined.

The rentier and welfare state syndromes are, however, deep and clear. There is hardly any private sector to speak of. Fishing off the reef is rare, and the pandanus trees and the palms that used to be major sources of food, and still are in poorer places like Kiribati, go unattended. Little is grown locally of fruit and vegetables. Instead virtually all food is imported. The eating habits of the Nauruans are none too healthy, and this, combined with a leisurely lifestyle, accounts for a high incidence of coronary diseases, diabetes and gout. The life expectancy of the Nauruans is surprisingly low, given their relative affluence and absence of hazardous occupations, only 55 years for males and 60 for females. Private initiative, which under normal circumstances is the engine of growth and economic progress, is blunted. The need to acquire skills and education is by many not perceived as important, which undermines the benefits of the free education system; truancy is reported to be a major problem. What little there is of private services, such as retail trading and restaurants, is mostly carried out by Chinese expatriates renting facilities from Nauruan landowners passing their copious spare time as best they can.[45]

Rentier lifestyle is neither wrong nor right in itself, although those who provide the rentier income by their own labor are prone to see things differently. A more fundamental question is whether the rentier income is sustainable. When it is based on mining a non-renewable resource it is not sustainable unless the rentier has followed the Hicksian rule of consuming

---

[44] Government of Nauru, General Information Handout, Statistics on Nauru.

[45] Much of the information in this and the following paragraphs is from Asian Development Bank (1998).

only the return on the permanent wealth and invested the remainder, to ensure that the wealth will in fact be permanent. How have the Nauruans fared in that respect?

To make their phosphate wealth permanent, the Nauruans faced the same basic options as any other society. Either they could have invested their phosphate rents in skills and industries which would be capable of maintaining or enhancing the previous standard of living once the phosphate runs out or they could have invested in financial wealth which would have permitted an even consumption profile over time. As our comments on the rentier welfare society indicate the former has not happened, and in all fairness the location of Nauru would hardly make that possible on the scale needed. That being said, the Nauruans could undoubtedly do a lot more than they do at present to help themselves; growing fruits and vegetables, raising battery chickens, and tourism, particularly reef fishing and diving, are activities that could take place on Nauru but which at present are nonexistent.

The Nauruans have indeed invested some of their phosphate money in trust funds, but mismanagement of these funds is one major reason why they do not suffice for maintaining the previous level of consumption. There are seven such trust funds, all of which have been managed by an umbrella organization located in Nauru House in Melbourne, the Nauru Phosphate Royalties Trust Fund. Some of these funds are owned by landowners on whose land phosphate has been mined while other funds are government funds. The value of these funds is believed to have fallen in the 1990s from a peak level of 1.3 billion Australian dollars to a paltry 225 million. At a rate of return of five percent, which is not an unrealistic assumption about the long term rate of return in financial markets, a fund of 1.3 billion dollars would have gone a long way towards being able to maintain the standard of living of the Nauruans. The sustainable rate of consumption from the trust funds is $r - g$, where r is the rate of return and g is the growth rate of the population.[46] At a rate of return of five percent the return on a fund of 1.3 billion would have been 65 million dollars. The growth rate of the Nauruan population is 2.7 percent, which leaves 2.3 percent of the fund to be spent sustainably. This would amount to 29.9 million dollars for a fund of 1.3 billion dollars, or roughly 3,700 dollars per capita for 8000 Nauruans. In 1991 the GDP of Nauru was estimated to be at 6,255 dollars per capita, but in 1996 it had fallen to 4,300 dollars per capita. The consumption sustainable

---

[46] The fund (X) per capita is X/N, where N is the number of people. If the rate of return on the fund is r and the rate of growth of the population is g, the fund at time t, if nothing were withdrawn, would be $X_0 e^{rt}/N_0 e^{gt}$, where 0 is some base year. Hence, the rate of growth per capita is $r - g$, so the amount that could be used every year without diminishing the per capita value of the fund is $(r - g)X_t$.

with the all time high fund of 1.3 billion dollars would thus have been less than the GDP in the crisis year of 1996, but not very much less.

This opportunity is now long gone. A fund of 225 million dollars would only provide a sustainable consumption of about 600 dollars per capita, using the same figures as above for rates of return and population growth and size. There are two principal reasons for why the funds have fallen so dismally. One is mismanagement. The funds have mostly been invested in real estate, which is often illiquid and which apparently has not yielded a good return. The Nauruans would have been better served by hiring professional broker houses for managing their funds on the basis of clear performance criteria, like the Alaska Permanent Fund and the Norwegian Petroleum Fund have done.

The other reason why the funds have dwindled is the fiscal position of the government of Nauru, which deteriorated rapidly in the 1990s. The dividends of the Nauru Phosphate Corporation, the main source of recurrent government revenue, dropped from an average of 22 million dollars per year in 1971-90 to five million in 1991, and to zero in 1993. The government expenditures, even if cut, far outpaced revenues, and the public debt is estimated to have increased from 275 million dollars in 1991/92 to 637 million dollars in 1997/98. To bridge the gap the government dipped into the phosphate trust funds. The funds' assets were used as collateral for taking up loans to meet the government deficit, which in effect diminished the value of the funds; the figure of 225 million quoted above is net of these loans. Furthermore, the Nauru government began to use the principal of the trust funds for paying for its expenditure, or borrowed money directly from the funds.

Yet another way in which the Nauru government tried to bridge the gap between its incomes and expenditures was borrowing from its own bank, the Bank of Nauru. Since Nauru uses the Australian dollar as currency, this had the effect of making the bank insolvent and thus unable to serve the citizens and businesses of Nauru. Checks drawn on the bank could not be redeemed for cash. It is instructive to compare this with what has happened in countries which have had their own national currency and which have tried, in a similar way, to use their own currency issuing bank for the same purpose. In those circumstances government borrowing has simply led to inflation, sometimes brief and unsustainable hyperinflations like the famous ones in Germany and Austria after World War I and in Hungary after World War II, and sometimes to protracted high inflation rates, like in Iceland, another small island nation, from the early 1940s to the late 1980s. The story of Nauru is a bit similar to the experience of the Faeroe Islands in the late 1980s where generous government subsidies to the fishing industry and irresponsible lending led to the collapse of the commercial banks in the islands in the wake of collapsing fish stocks.[47] The Faeroe Islands, being a part of the Danish state, use the Danish krone as currency.

Ironically, the Australian market, for which Nauru's phosphate was of vital importance for so long and for which the industry was initially developed, has become closed for Nauru at about the same time that the phosphate deposits are running out. The reason is that Nauru's phosphate has been found to contain too much cadmium, a contaminant for the soil. Instead Nauru has had to find markets elsewhere for its phosphates, which has meant lower prices and possibly smaller volumes in the short term. Needless to say, the closure of the Australian market has further aggravated the problems arising at the end of the phosphate era.

The future of Nauru

The non-sustainability of phosphate mining has thus made itself felt the harsh way. Clearly this eventuality has been insufficiently planned for, even if it was long foreseen. The funds that were set aside were insufficient or too badly managed for sustaining the consumption which the Nauruans had become accustomed to and so in violation of wealth management in the Hicksian spirit. Neither were other industries developed in Nauru which could have taken on the role of the phosphate mining as that era is coming to an end. To drive home the point of non-sustainability the phosphate mining has destroyed most of the interior of the island. What is left after the strip mining has run its course is a moonscape of rock pinnacles several yards high and useless for any purpose; a ghastly monument it might be said to short-sightedness and non-sustainability.

From a broader perspective the question of sustainability is less clear cut. The interior of the island of Nauru was never much used by the Nauruans, who used to live around the coast and were much fewer than they now are. Life in the old, pre-colonial days was doubtlessly sustainable on its own terms, but those terms could be harsh enough. Even then, in the absence of technological shocks such as those which came with the contact with Europeans, life was not a repetition of identical cycles or a continuity of an immutable process. The vagaries of nature could wreak havoc and pose stark challenges. A few decades ago the Polynesian island of Tikopia was hit by a devastating typhoon. According to custom, the chiefs assessed the number of people that the remaining resources would support. Having done this they would then start with the highest ranking families and count down to the cut off point, ordering everyone else to leave in a canoe and never return, which

---

[47] See Hannesson (1996).

meant near certain death.[48] How would modern societies with a less well entrenched hierarchy respond to a comparable disaster?

The contact with European colonial powers cannot be undone, and neither could it be avoided. Any solution to the problem of sustainability has to make allowance for that. Would the Nauruans have been better off if the phosphate industry would never have been developed? They would have been spared the present problems of adjustment, and to the extent adjustment means going without the goodies one has become accustomed to it can be difficult enough, and it could be argued that the Nauruans would have been better off never knowing about these goodies. The phosphate rents have brought the Nauruans things that their poorer cousins in, for example, Kiribati, know little or nothing about; a sanitation system, regular supplies of uncontaminated water, health services and education. Education and training, to the extent they are taken advantage of, should provide a greater mobility to greener pastures if the disappearance of the phosphate proves to be the modern day equivalent of a devastating typhoon. No one needs to depart these days in a canoe, but getting past the gatekeepers of the present day's affluent nations can certainly be a challenge.

The disappearance or even the deterioration of public amenities and services and in general the lowering of the standard of living that has accompanied the fall in the phosphate incomes, could certainly be called a crisis. But a crisis is also a challenge; it is in a state of crisis that people and nations change old habits that are no longer useful or perhaps directly counterproductive and get onto something new and better. What will the Nauruans do to cope with their present day crisis? Will they develop new ways to sustain their present standard of living and perhaps enhance it? There is talk of developing small scale services and industries for the home market. There is some potential for tourism, particularly for reef fishing and diving, and all those people who enjoy exposing their body to sunshine would find plenty of opportunity in Nauru. The island has an enormous economic zone where tuna and other fish can be caught, but the Nauruans do little of that at the present time. There is talk of developing new ways of phosphate mining, mine the phosphate that presumably is underneath the useless pinnacles and crush them to make a smooth surface on which one could build or spread new topsoil and grow things in the abundant sunshine, using desalinated water for irrigation. A comparison that comes to mind is the

---

[48] I'm indebted to Ward Goodenough, professor emeritus of social anthropology at the University of Pennsylvania, for this very interesting story. It was documented by Raymond Firth in his book *"Social Change in Tikopia,"* London: Allen & Unwin, 1959. He arrived on a boat just after the typhoon and, thanks to his arrival, the Solomon Islands government was radioed to send relief supplies, whose arrival obviated the necessity of sending people into exile.

island of Lanzarote where a small population used to subsist on precious drops of rainwater that fell in a few weeks of winter and irregularly at that. Now, thanks to mass tourism and help from the European Union, the island sustains a relatively large population and hordes of tourists on desalinated seawater and is a magnet for migrants from neighboring Africa. Optimists and pessimists about the human condition will do well to observe the development of Nauru, a miniature human laboratory, over the next few years to see whether the Nauruans grow with and out of their crisis or sink into a poverty trap from which they cannot escape.

CHAPTER FIVE
THE ALASKA PERMANENT FUND

"The Alaska Permanent Fund is a savings
account, ... which belongs to all the people of
Alaska. The beneficiaries ... are all present and
future generations of Alaskans."

Alaska Permanent Fund, Annual Report, **various
years.**

     Alaska has a colorful history. It was colonized by the Russians in the
late 1700s, mainly because of the fur trade. The United States bought Alaska
for 7.2 million dollars in 1867, a transaction that nearly cost the then
American Secretary of State, William Seward, his position. But Alaska
turned out to be good value for money, being rich in natural resources. It is
famous for its salmon fisheries, and one of the world's greatest ocean
fisheries, that for Alaska pollock, takes place in the waters off Alaska. And
Alaska is rich in minerals; Chaplin's famous "Gold Fever" was supposed to
take place in Alaska. Oil, in some ways the modern equivalent of gold, was
discovered in abundance on the north coast of Alaska in the late 1960s.
Alaska is the second newest state in the American Union; it became the 59th
state in January 1959, followed by Hawaii later that year.

     The oil fields on the north coast are large and were expected to be
highly profitable even at the low price of oil prevailing before the first oil
crisis in 1973. The Prudhoe Bay field, discovered in 1968, is the largest oil
field ever discovered in North America. When its production peaked in 1987
it accounted for 20 percent of the domestic production of crude oil in the
United States.[49] Most of the oil fields in Alaska are underneath land owned
by the state. Hence, the state gets revenue from oil as a landlord, both for
selling extraction licenses and as royalty on oil production. In addition there
is a severance tax on oil, and there is some revenue sharing between the state
and the federal government on oil produced from federal lands.

---

[49] University of Alaska, Anchorage, Institute of Social and Economic Research (ISER), Fiscal
Policy Papers, November 1992.

58

The auctioning of exploration licenses in the late 1960s brought the State of Alaska unexpected and unprecedented revenues. The Prudhoe Bay auction on September 10 1969, which dwarfed previous auctions, brought the state 900 million dollars and resulted in headlines of world war size in the newspapers. By comparison, the annual budget of the state in those days was less than 200 million dollars.

The state decided to use some of its oil revenue for furthering the economic development of Alaska. The development projects met with mixed success. One project paid prospective farmers for cutting down trees and growing barley on the land, but this was unsuccessful, and soon the Alaskan barley farmers were paid for replanting their land with trees. Another scheme of establishing dairy farms in Alaska also folded. Stories are told of a limited number of cows being marched repeatedly in front of project auditors to bolster the success of the project before its collapse, much as the soldiers of Tordenskjold.[50]

These unsuccessful attempts at government financed development policies were one reason why a different tack was taken once the revenues from oil production started to flow. Several meetings and public hearings were held on the subject of what to do with the oil money. Two American winners of the Nobel Memorial Prize in economics, Kenneth Arrow and Milton Friedman, were among those called upon to give advice. The proposal finally emerging out of this was to set up an investment fund to be financed by the oil money.

Before this proposal could be put into effect the constitution of Alaska had to be changed. The constitution explicitly prohibited earmarking of state revenues for specific purposes, with the exception of specific funds existing prior to the foundation of the State of Alaska in 1959. Needless to say there are valid reasons for avoiding the earmarking of incomes, particularly in a state with limited sources of revenue, as the case was in Alaska at that time. Needs change and incomes fluctuate, and there is a case for being able to respond to new needs or to sudden changes in revenue, which earmarking of revenue for specific purposes would hinder. It may be argued, however, that mineral rents are a special case because of their temporary character and the need to turn them into permanent wealth.

The necessary constitutional amendment was passed in a referendum with an overwhelming majority. According to the amendment a minimum of

[50] Tordenskjold (Thunder shield) was an officer in the Danish army who is reported to have persuaded the defenders of a fortress on the west coast of Sweden to surrender by marching his soldiers repeatedly in front of the fortress. The fortress was located in hilly terrain, and the soldiers were just about many enough for the first to have disappeared behind a hill and joined their comrades before the last became visible from the fortress. The defenders of the fortress were much impressed by the size of the army.

25 percent of the mineral royalties accruing to the state of Alaska must be deposited into the Alaska Permanent Fund, as the investment fund aptly came to be named. These 25 percent amount to a much lower share of the total oil revenues of the state, however, as revenues from severance taxes are excluded. These revenues have been of a magnitude similar to the royalty payments, so the statutory deposits into the Permanent Fund probably amount to 10 – 15 percent of the total income from oil.[51] The original proposal by the then Governor, Jay Hammond, had envisaged 50 percent of all oil revenue being deposited into the Permanent Fund, but the Alaska legislature watered that down by 75 percent, according to Hammond's own assessment.[52] The actual deposits into the fund have nevertheless been much higher, as the Alaska legislature has elected to make extraordinary deposits to the fund from time to time. The rest of the state's oil revenue goes into a "General Fund" which defrays the ordinary expenses of the state government.

As the name indicates, the Permanent Fund is intended to be permanent. All money that has been added to the fund's principal is protected by the constitution; it cannot be withdrawn unless the constitution is changed, which requires a referendum. This constitutional entrenchment of the Permanent Fund is probably an important guarantee for the preservation of the fund. Whether the constitutional entrenchment exists by design or default can be debated. As already explained, the constitution happened to prohibit the earmarking of money, an arrangement that long predates the enormous oil revenues. It is possible to view the constitutional entrenchment as a case of an institution, or a set of rules, coming about by chance but surviving because of their usefulness.

The use of the income earned by the fund is governed not by the constitutional amendment but by ordinary legislation. A part of the income is distributed to all inhabitants of the state as a "dividend" in the form of a check with an equal amount to each individual, after an application and a proof of residence for at least six months has been submitted. Originally the idea was to pay a premium to those who had lived in the state for many years, with the payment increasing with the length of residence in the state. This was challenged by a married couple of lawyers from New York who at the time were recent immigrants to the state. The premium for residence was finally struck down by the Supreme Court, which held it unconstitutional to discriminate among citizens in this way.

---

[51] According to the "Fall 1999 Revenue Source Book", Alaska Department of Revenue, the royalties paid to the Permanent Fund in fiscal year 1999 were about 14 percent of the total oil revenue of the state. See also "Fiscal Policy Papers," Institute of Social and Economic Research, University of Alaska, Anchorage, November 1992.

[52] Hammond (1994), p. 248.

The annual dividend is calculated so as to avoid large fluctuations from year to year that otherwise might arise due to variations in the rate of return on the fund. The formula is to use approximately one half of the average income for the last five years for dividends. A sufficient amount of the remaining income must be plowed back into the fund for "inflation proofing", that is, in order to preserve the real value of the fund. The remainder is kept in a reserve account to cover future deficits in case the income turns out to be insufficient to cover the dividend program and inflation proofing. This money can also be appropriated by the legislature, but it has not opted for using the money in the reserve account to defray general government expenses, instead adding it to the principal of the fund.

The dividend program was the brainchild of a previous Governor of the state, Jay Hammond. The purpose was to keep this money out of the hands of the politicians and to anchor the fund firmly among the Alaskan public. With an annual dividend from the fund the public was expected to develop an interest in maintaining the fund and prevent its erosion or outright liquidation. It can also be argued that using the return on the fund in this way rather than for lowering taxes or increasing public services is the most fair one, given that the oil is a resource for all Alaskans. Lowering taxes would not benefit those who do not pay any taxes, and not everyone is equally interested in or benefits equally from public services of various kinds. But most of Alaska's oil money has in fact been spent on paying for public services; the money that is left after the deposits into the Permanent Fund, and that is by far the largest share, goes into the General Fund which pays for public services. Ever since the beginning of oil production on the North Slope the State of Alaska has financed most of its current expenditure from the oil revenue. The state income tax was abolished in 1980 when the oil price was at its all time high in the wake of the second oil crisis. Twenty years later the fiscal position of the state had become such that the reintroduction of the income tax probably is the only alternative to using some of the income of the Permanent Fund to defray the expenditure of the state.

A fiscal bind

The Permanent Fund has been a successful institution so far. From its inception in 1976 it had grown to about 28 billion dollars as of September 2000. The return on the fund has grown handsomely over the years, and the way it has replaced the oil revenues of the state is almost a textbook example of how non-renewable resource wealth can be transformed into permanent wealth (see Figure 5.1). The dividend paid to each Alaskan has increased from 386 dollars in 1983 (there was a pent-up dividend of 1000 dollars in 1982, because of the lawsuit mentioned earlier) to 1963 dollars in 2000.

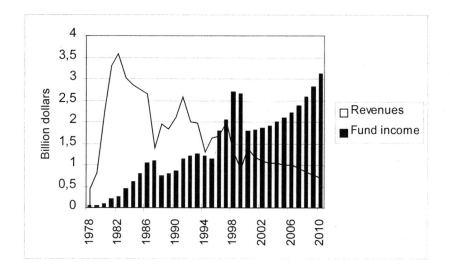

Figure 5.1. Oil revenues of the State of Alaska, except those earmarked for the Permanent Fund, and the income of the Permanent Fund. Sources: Revenues: *Fall 1999 Revenue Source Book*, Alaska Department of Revenue, Table K and Table 12 (predicted). Fund Income: *Alaska Permanent Fund, Annual Report 1999*, traced from a diagram.

The fund's investment policy is cautious, emphasizing security of returns over high but risky returns. The goal is to achieve a long term average real return of five percent but recently the fund has been doing much better, buoyed by the bull market on Wall Street. In the fiscal year ending on June 30, 2000, the rate of return was 9.18 percent, and the average annual return over a 16-year period ending in June 2000 was over ten percent.[53] Even after allowing for inflation, this is well above the long term goal of a real return of five percent. Originally most of the assets of the fund were invested in fixed income securities, but later the fund diversified into stocks, foreign as well as American. As of September 2000, 38 percent of the fund's assets were fixed income securities, 52 percent stocks, of which 15 percentage points were foreign stocks, and 9 percent was in real estate. These percentages have not changed much of late from year to year, but there has been a growing emphasis on stocks; in 2000 the fund trustees increased the allocation to stocks from 48 to 53 percent. A very small portion (around 1 percent) is invested in Alaska. With virtually all of its Permanent Fund

---

[53] Alaska Permanent Fund, Annual Report 2000.

62

invested outside the state, Alaska is in fact a small economy investing its mineral rents abroad, with the dividend essentially being rentier income.

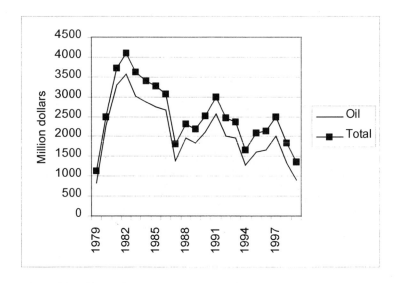

Figure 5.2. Unrestricted revenue of the State of Alaska, in total and from oil. Source: *Fall 1999 Revenue Source Book*, Alaska Department of Revenue, Table K.

While the Permanent Fund has been quite successful as an institution, Alaska's oil rents and their use have gotten the state entangled in major fiscal problems. The basic problem is oil revenues peaking early and then coming down gradually, or rather an inadequate response to this predictable development. Figure 5.2 shows the "unrestricted" revenues (i.e., excluding the mandatory deposits to the Permanent Fund) of the state of Alaska, in total and from oil. The revenues peaked at about 4 billion dollars in 1982 but were only 1.4 billion in 1999. The erosion of the real income of the state is even greater, as these figures take no account of inflation. The reason is the dwindling oil revenues; since the late 1970s oil has accounted for roughly 80 percent of the unrestricted revenue of the state.

This development has been met partly with reducing public expenditure in the state. Over the period 1991-98 the spending of the General Fund declined by 30 percent, but due to inflation and population growth the real spending per capita declined even more; if the real spending per capita had been maintained it would have been 3.3 billion dollars in 1998 instead of 2.3 billion, or more than 40 percent higher.[54] There is, needless to say, a limit

---

[54] See Goldsmith (1998).

to how far it is possible or at any rate desirable to go in this direction. Since the late 1970s the population of the state has grown from about 400,000 to 620,000 and with it the need for public expenditure on various services, not least education. Trimming expenditure has not been enough; the General Fund (the state budget) has shown a deficit since the late 1980s. In 1998 the deficit was more than 400 million dollars and about 1 billion dollars in 1999. The state has been able to cover the deficit through budgetary reserves accumulated around 1990 from dispute settlements with the oil companies. This is not, however, a renewable source of income; predictions in late 1999 were that these reserves would run out a few years into the new century.[55]

The decline in oil production was foreseen, even if it has been less rapid than predicted, due to the development of new oil fields and a higher recovery rate. That notwithstanding, the development of public expenditure in the state of Alaska has all the hallmarks of bad planning. Ideally public expenditure should be based on the need for services which it is deemed appropriate to have the government rather than the private sector provide. As incomes rise the need for infrastructure like roads and services like education increases rather than decreases, and even more so with a rising population. Given the extremely high share of the oil revenues in the total revenues of the state it could be argued that the need for a Permanent Fund, to be used as a source of income instead of the dwindling oil wealth, was particularly pressing. If Governor Hammond's ideas had been heeded and more of the oil revenues had been saved early on, the state would now find it much easier to finance its public services, provided it would be willing to use some of the return on the Permanent Fund for this purpose.

Given that it is not desirable to cut public expenditure by the same amount as the oil revenues go down, it is necessary to find new sources of revenue to replace dwindling oil revenues and budgetary reserves. The state of Alaska cannot print its own money, like some governments of sovereign states are prone to do when they are up against financial difficulties. Neither can it issue bonds, except for capital projects. This leaves the state with only two options for solving the deficit problem; either taxes will have to be raised and, with respect to the personal income tax, reintroduced, or the dividend program will have to be cut and the money used to defray the expenditure of the state. To an outsider, the choice Alaskans face does not look particularly onerous, as they pay virtually nothing for their government services. A study from the early 1990s showed that the average household in Alaska received more in dividends from the Permanent Fund than it paid in local taxes.[56] The

[55] Fall 1999 Revenue Sources Book, Alaska Department of Revenue, 1999.
[56] Fiscal Policy Papers No. 6, April 1991, Institute of Social and Economic Research, University of Alaska, Anchorage.

government services are thus essentially free for Alaskans, being paid for by petroleum revenues.

There are several arguments for using the income of the Permanent Fund for financing general government expenditure. One is that the Permanent Fund has made Alaskans richer than they would otherwise be, or perhaps more appropriately, made sure that the oil wealth discovered with much fanfare in the 1960s has not been entirely wasted on unsustainable consumption, private and public. One would expect that Alaskans would want to use some of this wealth for enhancing the level of public services. A logical conclusion following from that would be to use some of the income of the Permanent Fund for financing those services.

Another argument is that using the income of the fund rather than raising or reintroducing taxes would avoid the deadweight loss associated with taxes. Normally one may expect taxes to have undesirable side effects on the economy. Taxes on labor reduce the incentive to work and the willingness to employ people, taxes on capital income reduce the incentive to invest and retard economic growth, excise taxes reduce the amount of goods and services we buy and the utility we derive from them. Furthermore, the collection of taxes is not without cost. All of this would be avoided by using the income of the Permanent Fund rather than taxes to pay for public services. Lastly, Alaskans have to pay federal income tax on their dividends. If, instead of handing out the dividends, the return on the Permanent Fund were used for financing public expenditure in Alaska nothing would "leak out" of the state on the first round; Alaska would get a bigger bang for its buck.

There are, however, arguments for keeping the dividend program in whole or in part while relying on taxes to close the budgetary gap. One is essentially a fairness argument. The dividend program can be seen as an egalitarian scheme for income redistribution. Abolishing the dividend program rather than using taxes can thus be seen as a regressive form of taxation. Taxes are related to the level of income while the dividend is equal across the board. Abolishing the dividend program and using the income of the Permanent Fund to help pay for government services would hit low income families harder than doing so with an income tax. Reintroducing the personal income tax while keeping the dividend program might thus be a fairer way of financing public services and would also have the additional advantage of cementing the Permanent Fund. The tax rate, however, might have to be higher than what would be possible, or desirable, to achieve in Alaska. In one of his scenarios for the fiscal future of Alaska, Goldsmith (1999) calculates that a personal income tax would raise 350 million dollars per year. This is well below what would be necessary to maintain public services at the turn-of-the-millennium level.

Another argument for keeping the dividend and using taxes to pay for public services is an efficiency argument based on the deadweight loss of taxes but in the context of a less than an ideal world. The argument is that financing public services with taxes puts a necessary constraint on politicians; according to this argument it is the burden of taxes which prevents politicians from digging too deeply into the citizens' pockets. Behind this argument is an implicit assumption that politicians are predisposed towards overextending public services, as discussed in Chapter Three. Reluctance among the voters to pay taxes would in that scenario be a welcome if crude replacement for a cost of public services that no one observes or feels through his pocket book. Financing public expenditure from the income of the Permanent Fund, or from oil revenues for that matter, blunts the perception of the costs of these services. This is a likely reason why so much of mineral revenue is spent on current consumption as soon as it is earned and not transformed into permanent wealth, not just in Alaska but in many and perhaps most other places as well. Such expenditure gives immediate benefits to politicians and the electorate behind them, albeit at the expense of future generations.

Will the dividend program survive?

The option of using the Permanent Fund earnings to maintain the level of public services in Alaska has indeed been considered. In early 1999 the majority of the Alaskan legislature came to the conclusion that using some of the return on the Permanent Fund would be the appropriate way to go and put the issue to the electorate by way of an advisory referendum in September 1999. The debate that preceded the referendum was an interesting one to follow, even if doing so at a distance through the Internet hardly does justice to the color of the campaign. Using the earnings of the fund in this way would necessarily cut into the dividend program, and the forecast was a fall from an expected dividend of about 1700 dollars in the year 2000 to something like 1300 a few years hence. Unsurprisingly, perhaps, the voters would have none of that, and the proposal was soundly defeated by 83 percent of those who voted. Advocates of dipping into the dividend reminded the electorate that the original idea behind the Permanent Fund had been to secure a sustainable source of income for the state when the oil would run out. Opponents compared this to a poll tax where everyone is forced to pay the same contribution irrespective of income; the alternative to reducing the dividend, they said, was to reinstate the state income tax or raise other taxes which rise with income instead of falling flat on every man, woman and child. As so often in political campaigns there was the odd and unholy alliance, such as the professional protester with his guitar on an Anchorage street corner, who even forgot to apply for his dividend, campaigning against cutting the dividend together with the oil companies and other businesses unhappy at the prospect of having to pay more in taxes.

One interesting aspect of the debate about the dividend program before the referendum was how in the mind of many people the dividend has become an entitlement, much as the transfer payments associated with the welfare state (sickness benefits, child support, etc.). Many a Letter to the Editor from a supporter of the dividend program would make strong statements about "our oil", to which "we" are entitled, we being the residents of the State of Alaska. But on what is such entitlement based? The oil is located thousands of miles away from the population centers of Alaska. It takes advanced technology and costly equipment and some "roughnecks" to get it out of the ground, a process to which the ordinary Alaskan contributes nothing. Alaskans' ownership of the oil would not exist without the presence of state and federal government and the laws that state and federal politicians have enacted. The dividend program itself is a prime example of government interference and redistribution; no doubt the oil companies would rather avoid paying royalties and taxes but must yield to a strong state. Yet the dividend program seems to coexist peacefully with strong individualism and frontier spirit, if not anti-government phobia, in the minds of many Alaskans.

What, then, do we make of the dividend program? Is it the genie out of the bottle, making it politically impossible to use a reasonable share of the income of the Permanent Fund for defraying the costs of public services and thereby condemning Alaskans to private riches and public squalor? Or is it a fair and desirable income redistribution program, also fulfilling the dual role of cementing the Permanent Fund and protecting it from being raided by spendthrift politicians? Opinion on this is clearly divided in the state. However one answers these questions, the outcome of the referendum in September 1999 demonstrates the effectiveness of the dividend program as a guarantor of the Permanent Fund; if the dividend program goes out the window, the interest of the Alaskan public in the Permanent Fund and its preservation may follow suit. The Permanent Fund itself has demonstrated the usefulness of the investment fund approach as a device for turning non-renewable resource wealth into a permanent wealth. This is in stark contrast to the nearby Canadian province of Alberta, to which we are about to turn.

One argument that may be made against the dividend program is that it may attract immigrants to Alaska and so be self-defeating, from the point of view of those who already live in the state. So far, the dividend has clearly kept way ahead of population growth; having grown from 386 dollars per capita in 1983 to 1963 dollars per capita in 2000. One motivation for the original proposal of paying larger dividends to long time residents in the state was to avoid attracting residents purely for the sake of sharing in the dividend program. It is difficult to tell whether anything like that has in fact happened. The population of Alaska more than doubled from 1970 to 1999, growing from roughly 300,000 to almost 620,000. The population growth was most rapid in the 1970s and 80s, or more than 30 percent over each decade. This was the period when the pipeline transporting the North Shore oil across the state to the southern port of Valdez was built and much oil money was spent on developing various public services. The dividend program came into being in 1982 with the dividend rising gradually, so most of the immigrants in the 1970s and 80s were presumably attracted by work opportunities rather than a wish to share in the dividend. From 1990 to 1999 the population grew by only 12.6 percent, which is above the average for the United States as a whole, but many states experienced population growth above 20 percent in this decade; Arizona, Colorado, Georgia, Idaho, Nevada and Utah all did so, with Nevada on top with no less than a 50 percent increase. Since the dividend program has been much more generous in the 1990s than in the 1980s it does not appear to be a major attractor; in any event less so than the work opportunities in the 1970s and 80s. But Alaska is definitely one of the better-off states of the Union; in the years 1996-98 the median income of households in the state was the highest among all the 50 states, or slightly above 51,000 dollars, followed by New Jersey with just above 49,000. The lowest median household income among all the states was in West Virginia, just below 27,000.

CHAPTER SIX
THE ALBERTA HERITAGE FUND

"Are we prepared ... to put aside substantial
sums of current revenues from the sale of non
replaceable crude oil production, put it aside for
our grandchildren and not make it available for
current revenue needs; to use it for that day ...
when some of the wells may have gone dry ..?"

Peter Lougheed, premier of Alberta in 1976.[57]

If you look at it on a globe, Canada is a big country, one of the
biggest in the world. Most of it is, however, hardly inhabitable; the bulk of
the Canadian population lives within a couple of hundred miles or so from
the border with the United States. In a sense, therefore, Canada is a thin
stripe of land along a border stretching thousands of miles across the
American continent. Extraordinary historical circumstances are probably
needed to shape a country like that, and if it had not been for the might of the
British Empire in its heyday today's Canada would probably not have
existed. It seems highly appropriate, therefore, that Queen Victoria's
birthday is still celebrated in Canada.

Also, it seems understandable in that light that the names of some
Canadian provinces have a British imperial origin. The beautiful
westernmost province of British Columbia is one, Prince Edward Island in
the east is another. And then there is the province of Alberta, just east of the
Rocky Mountains, named after prince Albert who was married to Queen
Victoria. Canada was, however, not the only place in the British Empire
where states or provinces were named after members of the royal family.
Victoria is the name of one of the states of the Australian Commonwealth.
And the Australians in fact honored their queen twice; another state is named
Queensland, and there is little doubt whom they had in mind.

The finiteness of oil and gas wealth, amidst abundance of oil and gas
as such, is well illustrated by the case of Alberta. The province has deposits

---

[57] Quoted from Mumey and Ostermann (1990).

of oil and gas of good quality and location which give rise to rents of a considerable magnitude. In all probability these good quality deposits are quite limited, however, even if they still are not fully explored. In addition Alberta has abundant deposits of so-called tar sands, deposits of low quality hydrocarbons which need costly treatment for any usable oil to be wrung out of them. These deposits yield little or no rent; their development was initiated in the late 1970s when the price of oil was at an all time high and expected to remain so for a long time to come, but with a price of oil of less than 20 US-dollars, a value it has tended to fluctuate around for the last fifteen years or so, the production of oil from these tar sands has until recently not been profitable.[58] The case illustrates a point previously made about the need for a country, or a province as the case is here, to make the wealth embedded in good quality mineral deposits permanent. The country may have abundant minerals, but not necessarily deposits of good quality.

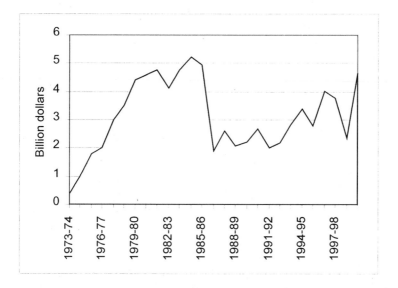

Figure 6.1. Alberta's non-renewable resource revenue. Sources: *Alberta Treasury* and *"Can we interest you in an $11 billion decision?"*, a brochure distributed in advance of a province-wide consultation on the future of the Heritage Fund, 1995.

Like the Alaska Permanent Fund, the Alberta Heritage Fund was established in 1976, in the wake of the energy crisis in 1973-74 which led to

---

[58] According to the Financial Times May 2, 2000, a special supplement on Alberta, the costs of producing oil from tar sands was about 20 Canadian dollars ten years earlier. Technological development has now reduced this to about 12 Canadian dollars or less.

the quadrupling of oil prices. The province of Alberta was by then well established as a producer of oil and gas[59] but its wealth was increased substantially as a result of the increase in the prices of oil and gas. Figure 6.1 shows the revenues of the provincial government from non-renewable resources, virtually all of which comes from oil and gas. These revenues increased about tenfold from 1974 to 1981, but fell by approximately 60 percent from 1985 to 1986, as a result of the fall in the price of oil. The revenues from oil and gas account for a substantial share of the revenues of the provincial government; in the early 1980s this share was about 40 percent but dropped as a result of the fall in the oil price in 1986 and has fluctuated around 20 percent since then (see Figure 6.2).

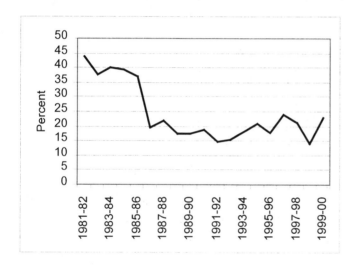

Figure 6.2. Share of the total revenue of the provincial government of Alberta coming from oil and gas. Source: *Alberta Treasury.*

The provincial government gets its revenue both through taxes and as royalties; about 80 percent of the production of oil and gas in Alberta takes place on Crown land,[60] over which the province has jurisdiction. Largely due to the oil and gas revenues, Alberta is the richest province in Canada. The gross domestic product of the province was just above 36,000 Canadian dollars[61] per capita in 1998, followed by Ontario with a little less than 33,000,

---

[59] Oil was discovered in Alberta in 1947.
[60] Figure quoted from Warrack and Keddie (undated).
[61] All values in this chapter are in Canadian dollars unless otherwise stated.

while Newfoundland and Prince Edward Island are the poorest provinces in the Canadian federation, with a GDP per capita of about 21,000 dollars.[62]

As the name suggests, the Heritage Fund was established to take care of the province's heritage, the non-renewable resource wealth, which in practice means oil and gas. Taking care of this heritage would seem to amount to making the mineral wealth permanent through investing the oil and gas rents. It is not clear, however, whether it was wealth management in this sense or simply postponing the use of the oil money to a later date which was the purpose of the fund. The explicit goals set for the fund, and the documents and speeches outlining its purpose, contain elements of both. In his introduction of the legislative proposal for the Heritage Fund, the premier of Alberta at the time, Peter Lougheed, outlined four objectives for the fund:[63]

(i)   Act as a future source of revenue, either through income from the fund or from the fund itself.
(ii)   Reduce the debt load that might develop at some future and perhaps not very distant point in time.
(iii)   Improve the quality of life in the province.
(iv)   Strengthen and diversify the economy of the province.

As the first point makes clear, indefinite wealth preservation was not the single objective; the depletion of the fund at some future date clearly was seen as an option. Hence the fund was not seen exclusively as an instrument for making the oil and gas wealth permanent but also as a displacement in time of using up the petroleum money, in the spirit of don't chew on more at a time than you can swallow. The fourth point could, however, be interpreted as making the petroleum wealth permanent, as strengthening and diversifying the economy of the province would require productive investments of some kind. So, already at the outset, the long term wealth management perspective was much weaker for the Heritage Fund than for the Alaska Permanent Fund. The principal of the latter, it will be recalled, is protected by the constitution of Alaska.

Initially the savings plan of the Heritage Fund was considerably more ambitious than that of the Alaska Permanent Fund. At the outset, 30 percent of the province's oil and gas revenues were deposited into the Heritage Fund

---

[62] Calculated from a database maintained by "Computing in the Humanities and Social Sciences," The University of Toronto. The differences in income per capita are less than this, because there is substantial redistribution of income between the provinces through various transfer payments of the national government.
[63] Quoted from Mumey and Ostermann (1990); see also Smith (1991) on the purpose of the fund.

while the statutory deposits into the Alaska Permanent Fund amount to half of that or less. As oil and gas prices started to slip in the early 1980s and the American recession spread to Canada this percentage was lowered to 15, and after the precipitous fall in the oil price in 1986 the deposits into the fund ceased altogether. This merely reflected the fiscal situation in the province; government revenues from oil and gas fell from about five billion dollars in 1985-6 to less than two billion dollars in 1987 (see Figure 6.1), and the provincial government began a period of substantial budget deficits and accumulation of debt (see Figure 6.3). Clearly it would not have made much sense to put money into the fund at the same time as the province was accumulating debt; any real buildup of financial wealth requires budgetary surpluses. The Alberta government and legislature may, however, be criticized for taking an insufficient interest in preserving the petroleum wealth.[64] The provincial taxes in Alberta have been and still are lower than in other Canadian provinces while public expenditure is higher. Preserving the petroleum wealth would have demanded an increase in taxes or cuts in expenditure, or both, in order to produce the budgetary surplus required for making any real savings from the petroleum revenue.

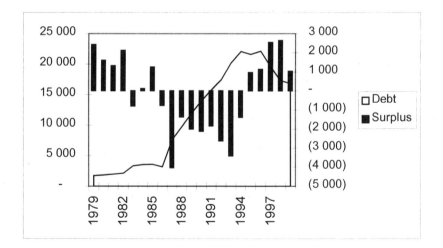

Figure 6.3. Budget surplus (bars, right scale) and debt (area, left scale) of the provincial government of Alberta, in millions of dollars. Source: *Alberta Treasury.*

---

[64] See Scarfe and Powrie (1980), p. 174-5.

Somewhat surprisingly, the Heritage Fund was left virtually intact while the provincial government continued to accumulate debt. The fund reached a maximum of 12.7 billion dollars in 1987, but since no more deposits were made and the fund income was used for financing ordinary government expenditure the real value of the fund eroded under the force of inflation. One reason why the fund was not used for reducing the provincial debt may have been that its assets were not liquid enough. The losses involved in not using the fund to pay off the provincial debt were also limited. Normally governments have to pay a higher interest on their debt than they earn on their financial assets but in some years the fund earned a higher return than the government had to pay as interest on its loans.[65]

How the fund was managed

In contrast with the Alaska Permanent Fund, which is run by an independent corporation at an arm's length from the state government and legislature, the Alberta Heritage Fund was run by the Alberta Treasury and at the discretion of the provincial government. The provincial parliament has the power of overseeing, and there is a standing parliamentary committee which reviews the management of the fund. The effectiveness of this has been called into question. The committee's reports on the fund have at times become politicized along government versus opposition lines.[66]

Prior to its restructuring in 1997 the Heritage Fund was organized in four divisions. The Commercial Investment Division was expected to earn a commercial return and invested in shares and bonds. This operation appears to have met with a success comparable to the results of other investment funds in Canada.[67] The Canada Investment Division lent money to other Canadian provinces. These loans came to an end in 1982, and on the present repayment schedule they will have been repaid in 2005. Even if these loans were made at concessionary rates they yield presently a high rate of interest, having been issued at a time of high inflation and high interest rates.

The Capital Projects Division invested in infrastructure, education and research. These investments yield no financial return, and the assets have no market value but nevertheless used to be listed in the fund's accounts at a deemed value. Even if there is no financial return or market value of assets this does not mean that these expenditures have been in vain, but no

---

[65] *"Future Directions for Alberta's Heritage Fund"*, Report of the Alberta Heritage Savings Trust Review Committee, March 28, 1995, p. 2.

[66] On the review committee during the first years of the Heritage Fund, see Pratt and Tupper (1980).

[67] See Mumey and Ostermann (1990).

analysis that we know of has been made of the success or otherwise of these expenditures.

Finally there was the Alberta Investment Division. In 1988 over a half of the fund's financial assets were held by this division, but by 1996 that share had fallen to 20 percent.[68] The bulk of this was invested in various Crown corporations; the Alberta Mortgage and Housing Corporation, Alberta Agricultural Development Corporation, Alberta Government Telephones Commission, Alberta Municipal Financing Corporation, and Alberta Opportunity Company. These Crown corporations were for the most part not a financial success. The "competitive" interest they paid on the loans from the Heritage Fund was in part financed by grants from the provincial government itself, making these incomes of the fund illusory. Correcting for this, Mumey and Ostermann (1990) found that the real rate of return for this division was about two percent per year for the period 1981-88, as against ten percent for the Commercial Division. Apparently the government of Alberta used the fund to invest in undertakings in the province which were not profitable enough to attract investment funds from other sources. One of the least successful undertakings was a loan of 120 million dollars to the company Millar Western in 1987; this was sold for 28 million dollars in 1997.

In financial terms the fund was thus a mixed success. The lack of financial success in areas where that criterion is not applicable (the Capital Projects Division) is not the issue here, but the lack of success of the Crown corporations is, since these were commercial undertakings which should have earned a competitive return. The history of the fund casts an unfavorable light on having governments run development banks financed from tax money. In such arrangements political expediency tends to overshadow financial objectives, and even if the latter may at times be insufficient as criteria for success in economic development, political criteria are not likely to be any better.

The organization of the Heritage Fund reflected its various goals. The Alberta Investment Division and the Capital Projects Division were responsible for transforming the petroleum wealth into other forms of wealth in the province, real as well as intangible. The success with which this was done is another issue, and in a number of cases it does not seem to have been done very successfully. It is tempting to seek the explanation for this in the fact that the fund was directly controlled by the Treasury instead of being an autonomous institution like the Alaska Permanent Fund. Investing in Crown corporations providing mortgages to urban dwellers, and telephones and

---

[68] The 1988 figure is from Mumey and Ostermann (1980) and the 1996 figure from the Alberta Heritage Savings Trust Fund Business Plan 1997.

76

irrigation in rural Alberta, may have paid off in political benefits what was lacking in commercial return. Foregoing commercial rates of return was at least in some cases an explicit decision. One of the original goals of the fund was "strengthening and diversifying" the Alberta economy. This was changed in 1979 by the legislature to "strengthening or diversifying".

The Canada Investment Division can be seen as a result of Alberta's position within the Canadian federation as a wealthy province with a relatively small population, selling its high priced oil and gas to the rest of the country.[69] There was considerable resentment in the rest of Canada towards Alberta becoming rich at the expense of the rest of the country through an exorbitant rise in the price of oil and gas. The government of Canada had for a number of years in the late 1970s and early 80s a national energy policy which meant that the Canadian domestic prices of oil and gas were way below the world prices.[70] This, in effect, amounted to a major transfer from Alberta to Canadian consumers, mainly located in central Canada, compared to a situation with oil and gas being sold at world prices.[71] The Canadian Investment Division of the Heritage Fund provided loans to other Canadian provinces at concessionary rates, the purpose of which most likely was to seek goodwill among other Canadians through letting some of the oil money benefit other provinces.[72]

The subject of the position of Alberta within the Canadian federation raises again the question to what extent it is possible for a single, resource rich province to invest its mineral wealth for the benefit of its own inhabitants. In Canada, as in Alaska, there are no formal restrictions on the movement of people across provincial boundaries. If a province like Alberta tries to achieve a considerably higher income for its inhabitants than people obtain in the rest of the country, one would expect this to attract migrants from other provinces, which would dilute the wealth enhancing effect for the inhabitants previously in the province. There are, as already stated, considerable differences in GDP per capita among the Canadian provinces, even if the differences in incomes are much less, due to transfers of various kinds. To some extent people seem to migrate in response to these differences; since 1971 the population of Alberta has almost doubled; it has grown from 1.7 million in 1971 to 3.0 million in 1999 while that of

---

[69] Alberta's population is 3 million while there are 30 million in all of Canada (figures from 1999).
[70] According to Courchene and Melvin (1980), p. 204, the Canadian domestic price of oil in 1979 was just about one-half of the world price.
[71] See Smith (1980), pp. 143 – 144.
[72] The efficacy of this has been doubted; see Stevenson (1980), p. 273.

Newfoundland, for example, has hardly grown at all.[73] The province of British Columbia has, however, had a development similar to Alberta despite being poorer; its population has grown from 2.2 million in 1971 to 4 million in 1999. The GDP per capita of British Columbia was just below 28,000 dollars in 1998, compared to just over 36,000 dollars in Alberta.

The new Heritage Fund

Until 1982 the income of the Heritage Fund was reinvested in the fund, but after that it was used to defray the expenditures of the provincial government. The fund stopped growing after 1986 and while the nominal value of the fund declined only slightly its real value was eroded through inflation. By the time the reorganization of the fund was initiated (1995) its nominal value was down to 11.4 billion dollars.[74] By the mid-1990s the wisdom of preserving the fund while the province was wading ever deeper into debt was increasingly being questioned. A province-wide consultation with the voters was initiated on what to do with the fund, whether to preserve it or to liquidate it and use it for reducing the provincial debt. In the end it was decided to keep the fund. Starting in 1997 the portfolio of the fund is to be transformed, over a ten year period, into one of blue chip bonds and stocks. The fund will thus gradually become somewhat similar to the Alaska Permanent Fund in that it will invest in low risk financial assets.

Another parallel with the Alaska Permanent Fund is that a sufficient part of the fund income must now be retained in the fund for inflation proofing. The remaining income can be used for defraying the costs of the provincial government. In the fiscal year 1999-2000 230 million dollars of a total fund income of 1,169 million dollars, or roughly 20 percent, was retained in the fund, with the rest contributing to paying general expenses of the provincial government. There are no immediate plans to revive the fund as a repository of oil and gas rents to be invested, so its wealth transformation role is one that belongs to the past; its main role appears to be one of a budgetary reserve, a savings account for a rainy day. At the end of the fiscal year 1999-2000 the fund was worth 12.3 billion dollars.

The preservation and structural change of the Alberta Heritage Fund is a part of a political reorientation in the province towards paying off the provincial debt. In fact, without such reorientation it would not have made

---

[73] The population of Newfoundland was 530,000 in 1971 and reached a peak of 580,000 in 1992, but has since fallen to 541,000 in 1999.

[74] As assessed by dealers, according to the brochure distributed to all households in Alberta as a part of a province-wide consultation on what to do with the Heritage Fund. Later annual reports of the fund show a figure of 11.8 billion dollars.

much sense to keep the fund. The provincial deficit turned into surplus in the fiscal year 1994-95 (see Figure 6.3), and by 1999 the net debt had been eliminated. The plan now is to eliminate the provincial debt entirely over 25 years, according to The Fiscal Responsibility Act of March 1999. The restructured fund will be an ordinary financial investment fund with a mandate to maximize its financial return, with due account taken of risk. The objective of the fund is now stated as follows: "to provide prudent stewardship of the savings from Alberta's non-renewable resources by providing the greatest financial returns for current and future generations of Albertans".[75]

Compared to the Alaska Permanent Fund the Alberta Heritage Fund must be regarded as a failure. The process of transforming non-renewable resource wealth into permanent wealth has long since come to an end, and such transformation as was attempted was less than fully successful. Why did things turn out the way they did? Perhaps the fiscal needs of the province were greater than those of Alaska at the critical point of falling oil prices in 1986, perhaps the willingness to prune unsustainable public consumption was less, both on behalf of politicians and the electorate at large. But the institutional structure was also different and, for the Heritage Fund, less conducive towards long term preservation of wealth. The Alaska Permanent Fund is an institution at an arm's length from the government and the legislature, with a clear financial objective. The Heritage Fund is a government institution and, prior to the 1997 revision, with several and sometimes inconsistent goals; earning competitive returns on financial investments, but also province building with intangible benefits or low financial returns. More importantly perhaps, the Heritage Fund had no dividend program or any other mechanism which might have anchored it in the minds of the general public and secured support for an investment policy that was successful in financial terms. Probably because of this the interest among Albertans in the Heritage Fund appears to have been minimal. Asked about the public debate and mail referendum on the future of the fund, introduced in 1995 with some fanfare, a Calgary academic had to think long and hard about when he moved to the province, to find out whether there was any reason to expect him to remember the episode. It finally came to him: "oh yes, I remember that letter, I threw it into the garbage can."

---

[75] Alberta Heritage Savings Trust Fund Business Plan 2000-03.

CHAPTER SEVEN
THE NORWEGIAN PETROLEUM FUND

"The desire to separate the earning of the
petroleum revenues from their use has, in the
view of this member, been given too much
emphasis in the conclusions of the committee.
The objective is to utilize the petroleum
revenues."

**Tora Houg, member of the "Extraction Rate
Committee" (Tempoutvalget), NOU 1983:27, p.
100 (Author's translation).**

By the late 1990s Norway had become the second largest oil exporter
in the world, after Saudi Arabia. Its history as a major oil producer is
nevertheless short. The beginnings were inauspicious. Over a few years in
the 1960s the expectations with regard to oil production in the Norwegian
part of the North Sea went from deep pessimism to optimism and euphoria.
As recently as the late 1950s Norwegian geologists regarded the probability
of finding oil underneath the North Sea as negligible, a view that only slowly
started to be revised after the discovery of the Groningen gas field in the
Netherlands in 1959.[76]

The first approach to the Norwegian government regarding
exploration of the North Sea continental shelf came as a surprise in 1962.
Slowly the necessary legal framework was put in place, and in 1966
exploration started, but many years went by without any profitable finds
being made. In late 1969 Phillips Petroleum, the pioneer on the Norwegian
shelf, had to abandon an unfinished well in the southernmost tip of the
Norwegian sector. Phillips had had enough of dry holes and, despairing of
success, asked permission to send its men home for Christmas. The
Norwegian authorities insisted that the exploration commitment be carried

---

[76] A letter from the Norwegian Institute of Geology ("Norges geologiske undersøkelser") to
the Ministry of Foreign Affairs in February 1958 stated that the probability of finding coal, oil
or sulfur on the continental shelf off the Norwegian coast was insignificant. Quoted in
Erlandsen (1982), p. 80.

out according to plan. The day before Christmas Eve 1969 oil was struck on what later became known as the Ekofisk field. The development of the field took several years, so the beginnings of oil production in Norway coincided with the first energy crisis and the quadrupling of oil prices. The oil era thus started out on a note of optimism, which was further fueled by continued discoveries and a new oil price hike after the revolution in Iran in 1979.

It quickly became clear that the petroleum industry would be a major boost for the Norwegian economy. The downside was that this would most likely cause serious problems of structural adjustment; old and established industries would have to give way for the new industry and other activities that were likely to grow in its wake. Government papers on oil policy published in the early 1970s saw the absorption of the oil revenues as the main challenge. Most of the good things the public wanted realized with the oil money required human and other domestic resources in short supply, even if some needs could be met by imports. As one government official is reported to have remarked off the cuff: "people need to realize that it's oil and not nurses that we're pumping up from the North Sea." But ever since the beginnings of the oil era the question how the petroleum wealth could be made permanent has been curiously absent in government papers on petroleum policy in Norway despite their generally sober tone and realistic outlook. Wealth management has almost exclusively been seen in the perspective of postponing the use of the petroleum revenues until such time as they could be absorbed into the domestic economy and not from the perspective of turning them into permanent wealth.

The absorption problem was again a central theme in a report on the appropriate rate of petroleum extraction published in 1983 (NOU, 1983). The report was produced by a committee appointed by the Norwegian parliament, with a mandate to inquire into the appropriate rate of extraction. A central tenet of the report was that the rate of extraction should be determined with a reference to the needs of the country and its capacity to utilize the oil and gas revenues. This essentially amounts to saving the petroleum wealth underground, an investment policy which, in the light of the development of the price of oil, is not obviously profitable. The idea of saving some of the oil revenue in a fund to be invested abroad was brought to the fore in this paper, but more for the purpose of acting as a buffer against difficult times, or to defer income that could not be spent immediately, rather than as a vehicle to transform transient wealth into permanent wealth. One member of the committee, also a member of parliament, found it necessary to voice a slightly dissenting opinion in response to the fund idea, to the effect that ultimately the oil incomes must be used in the country.[77] The wealth

---

[77] See quotation at the beginning of this chapter.

management idea does not appear, therefore, to have gone much beyond shifting the income profile over time; there was little if any trace of the idea that oil rents could or should be transformed into permanent assets.

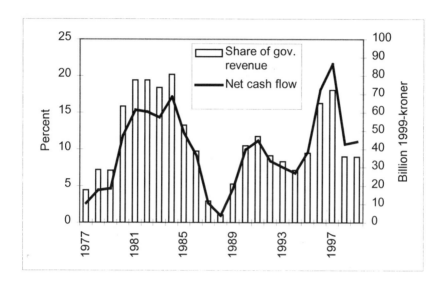

Figure 7.1. Government revenue from oil and gas extraction in Norway. Source: Ministry of Oil and Energy, *"Faktahefte"*, and Statistics Norway, *Yearbook of Statistics* (various years) and *Historical Statistics 1994*.

The oil price fall in 1986 made any plans for a petroleum fund next to irrelevant. Most of the oil rent vanished over night. With prices recovering, rent and tax income recovered as well, but the government was heavily engaged as a passive partner in several projects on the continental shelf; its net cash flow from oil and gas dipped almost to zero in the years 1987-89 (see Figure 7.1). Nevertheless, the Government Petroleum Fund (Statens Petroleumsfond) was established by law in 1990. Even if the petroleum revenues recovered, amounting to about ten percent of government revenue in 1990-92, no deposits into the fund were made until 1995; the economy was in severe recession in 1990-92 and most of the oil money was spent to fight the recession.

How the fund works

The Norwegian Petroleum Fund differs from its counterparts in Alaska and Alberta in several important respects. The amount to be invested each year is not specified, it is simply whatever budget surplus the Norwegian parliament may pass each year. In theory all government revenue from petroleum taxes and equity sharing arrangements on the continental shelf goes into the Petroleum Fund, but the amount needed to balance the budget is automatically withdrawn, with the remainder being the annual deposit into the fund. The buildup of the fund is thus entirely at the mercy of the parliamentary majority at each point in time, and that same majority could deplete the fund by running a deficit on the government budget for a sufficiently long time.

This arrangement may seem more than a little too haphazard to serve as an instrument for wealth management. It bears more than a superficial resemblance to a person who keeps a reserve account in the bank in addition to his checking account and withdraws money from the reserve account in case the checking account is out of balance. Government officials and politicians responsible for framing the rules for the fund did not see it that way, however. They saw it as a useful pedagogical device for parliamentarians, forcing them to take an overview of government finances in their entirety. Both the Alberta Heritage Fund and the Alaska Permanent Fund illustrate that this argument is not without merit. The state of Alaska has been running a deficit for years, drawing on reserves that soon will run out and in effect reducing its financial wealth on one account (the budgetary reserve) while building up the Permanent Fund. The Alberta Heritage Fund became in effect a fiction as the province for years had a budget deficit and piled up debt. Nothing much will be accomplished in terms of investment if the government accumulates debt at the same time as it puts money aside in an earmarked account.

The efficacy of this pedagogical device largely remains to be tested. Ironically, both "too low" and "too high" oil prices may make it difficult for parliamentarians to build up, or maintain, the Petroleum Fund. What will they do, for example, if the oil price again drops like a stone like in 1986 and the Norwegian economy goes into recession, as it did a couple of years later? In late 1998 when the oil price reached a low point of less than 10 dollars per barrel it looked for a while like that acid test was coming. The fall in the oil price caused a major structural upheaval in the Norwegian petroleum industry, with one of the Norwegian oil companies being gobbled up by the other two and with investment in the oil industry declining dramatically. The crisis was, however, short lived, except for the dearth of investment in oil field development; by mid-year 1999 the oil price had recovered to about 22 dollars per barrel.

But an exceptionally high oil price and the resulting high rents could also make it difficult to put aside money in the fund. As the oil price rose from the frightening depths attained in late 1998 to 30 dollars per barrel or more by the summer of 2000 the Norwegian debate took a clear and perhaps predictable turn. Examples were cited of sore needs in the public sector; badly maintained school buildings, long waiting lines for medical care, low salaries and recruitment problems for nurses and teachers, etc. No matter how the story began the punch line was always the same, how can such needs go unsatisfied in a country that is virtually drowning in cash? It is probably no exaggeration to say that a large part, perhaps the majority, of the Norwegian electorate does not at all see the point in salting away the oil money in foreign lands while immediate needs go unfulfilled. At any rate it is a point frequently made by journalists and others that such an apparent paradox cannot be explained convincingly to the general public. And when trying to explain the apparent paradox, most politicians point to absorption problems, i.e., inflation being fueled by spending more of the oil money immediately, rather than invoking the need, or the desirability, to preserve the petroleum wealth for coming generations.

The apparently limited appreciation among the general public of preserving the petroleum wealth for future generations raises the question of what incentive mechanisms would be needed for generating such interest. To accomplish this it is probably necessary to give each individual or household a stake in the fund such that he or she benefits directly from building up and preserving the fund and managing it well. The dividend program of the Alaska Permanent Fund, already discussed, is one such mechanism. But there are other possibilities. One is to turn the petroleum fund into a pension fund. The Norwegian retirement system is largely of the pay as you go type; i.e., current contributions to the retirement scheme pay for the pensions of those who have already retired, as the case is in many other European countries. As the number of pensioners rises in relation to those of working age this will become increasingly difficult, a problem shared with many other European countries. The issue has indeed been raised how the Petroleum Fund would help pay for the increasingly heavy retirement burden.

The Petroleum Fund will not be a long term solution, however, unless the oil wealth is preserved and made permanent, using the return on the fund rather than the fund itself to pay for the retirement. Making old age pensions conditional on the financial health of the Petroleum Fund would contribute to preserving the fund by giving each generation a stake in the fund. Better still might be to individualize the fund; i.e., give each individual a share in the fund and make its preservation his or her personal

84

responsibility within reasonable limits, with the risk of otherwise taking a cut in the retirement income. Earmarking the return on the fund for pensions would be less unjust from an intergenerational perspective than it may seem, as the pensions would otherwise have to be financed by the working citizens, to the extent the real values of pensions would be maintained. Such a fund, if preserved, would be a "bounty" for each generation, to be enjoyed at the end of the day rather than at high noon.

The issue how the Petroleum Fund could alleviate the future retirement burden was recently addressed in a special supplement to the National Budget.[78] Curiously, this special supplement illustrates better than perhaps anything else the aforementioned absence of a long term wealth management perspective in most government papers on petroleum policy. The supplement focussed on the long term fiscal dilemma of declining incomes from the petroleum industry and a rising burden of retirement payments. Alternative forecasts were made for the fiscal position of the Norwegian government for the next half century, for different assumptions about the revenue flow from petroleum. In all alternatives the Petroleum Fund peaked some time around 2020 and then declined, vanishing some time between 2030 and 2050. The option of making the fund permanent and using its income to help defraying government expenditures including retirement payments might seem to suggest itself but was in fact not even mentioned.

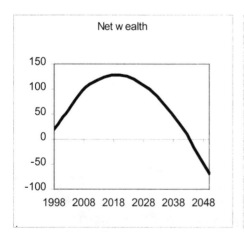 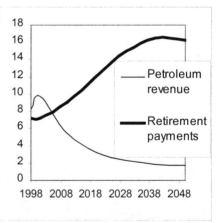

Figure 7.2. Projections of petroleum revenue, retirement payments, and net financial wealth (Petroleum Fund) of the government of Norway, in percent of GDP. Source: *Stortingsmelding* (Report to Parliament) No. 1, 1997-98, Table 3, p. 192.

---

[78] Stortingsmelding (Report to Parliament) No. 1, 1997-1998, "Nasjonalbudsjettet 1998".

This is all the more curious as the gist of the whole exercise was the increasing long term fiscal burden due to retirement payments. These payments were predicted to stabilize after 2040 but not to decline and still less to vanish altogether, as the Petroleum Fund. There was no indication in the supplement of how the continuing burden of retirement payments was supposed to be dealt with after the fund has vanished. The fact that these events lie far into the future is hardly a good reason to dodge the issue; although we cannot predict the age structure of the population half a century from now very well we can be sure that the retirement burden will not disappear around that time. Figure 7.2, constructed from the supplement, shows the development of the retirement payments, income from the petroleum sector, and the Petroleum Fund.

Are the savings sufficient?

Despite the lack of interest among the general public, the savings of the oil money have been quite high since the government began to invest in the Petroleum Fund in the mid-1990s. Well over one half of the government's net cash flow from oil has been saved in the fund annually since 1996 (see Table 7.1). These savings have been variable, however; they were close to three quarters in the boom year of 1997 but only two thirds in the bust year of 1998 when the oil price was low and the Asian Crisis in full bloom. How does this compare with what would be needed, according to the Hicksian permanent income criterion? In the government budget for 2000 the government's share of the petroleum wealth was estimated at 1830 billion kroner.[79] This is the present value of the expected net cash flow accruing to the government, discounted at 4 percent per year. If this wealth is to be preserved in per capita terms, the maximum amount that could be consumed annually is 3.4 percent, or 62.2 billion kroner. The 3.4 percent is the difference between the expected rate of return on the wealth, which is the 4 percent used to discount the expected cash flow, and the growth rate of the population, which in the 1990s has been about 0.6 percent on an annual basis. The expected net cash flow in 2000 was 138.1 billion kroner, so the minimum amount that had to be saved was 75.9 billion, assuming that the return on the Petroleum Fund will be saved in the fund and that no additional savings are necessary to make up for possible "sins" of insufficient savings in the past. The savings of the oil money budgeted for 2000 were well beyond this. The budgeted deficit of the government, excluding the oil money, was

---

[79] Stortingsmelding (Report to Parliament) No. 2, 1999-2000; "Revidert nasjonalbudsjett 2000". The numbers in this and the following paragraph are taken from this publication.

86

13.8 billion kroner, so the budgeted savings of the oil money were 124.3 billion kroner (138.1 less 13.8), about half again as high as the minimum savings needed to satisfy the Hicksian criterion.

Table 7.1. The Petroleum Fund: deposits, fund balance, and government net cash flow (NCF) from oil and gas, in billion "kroner". Sources: *Stortingsmelding* (Report to Parliament) No. 1, various years, and Bank of Norway. Figures for 2000 are estimates from the National Budget for 2001.

	1995	1996	1997	1998	1999	2000
Deposits	1.2	44.2	64.0	28.0	26.1	142.3
NCF from oil and gas	38.5	69.9	86.8	45.0	44.6	160.2
Deposits in % of NCF	3	63	74	62	59	89
Fund balance, end of year	1.2	46.3	115	172	221	385.1
Rate of return on fund (%)				15.7	13.9	

It may be argued that more than this minimum should be saved, because of the increasing future retirement burden. So-called generational accounting has indicated a need to save 5-20 billion kroner per year to avoid a future rise in taxes to finance the retirement burden.[80] Adding this to the previously calculated savings needed to preserve the petroleum wealth per capita still lands us well below the budgeted investment of the oil money for 2000. Possibly it may also be argued that the savings policy is overcautious because of unduly pessimistic assumptions about the price of oil. The assumptions made about the future price of oil in calculating the oil wealth were as follows: 23.75 dollars per barrel for the year 2000, 18.25 dollars for 2001, and 15.6 after that.[81] In the summer of 2000 the oil price shot above 30 dollars per barrel, but needless to say it remains to be seen whether the world has entered a period of permanently high oil prices. Such expectations have been proven wrong before, and as the reader will recall the oil price briefly went below ten dollars per barrel in late 1998 and early 1999.

As to the investment policy, the Petroleum Fund is entirely invested abroad. This is in part due to a desire to limit the demand pressure in the domestic economy, and in part an attempt to diversify away from oil, on which the entire economy and the government in particular depend heavily. The assets are invested in government bonds and company shares in various parts of the world. The guidelines specify that 30 – 50 percent of the assets can be invested in stocks and the rest in bonds. The fund is not supposed to

[80] See *Gjensidige NOR Spesialrapport*, July 3, 2000.
[81] The figures in the government budget are expressed in kroner. These have been converted to dollars by applying an exchange rate of 8 kroner per dollar, which is close to the rate in late 1999.

be an active owner, and there is an upper limit of one percent of total stock that the fund can own in any single company. The fund is not an independent institution but is owned by the government and managed by the Bank of Norway. The Bank has elected to engage a number of foreign brokers to carry out the day to day management of the portfolio, with a mandate to do no worse than an agreed reference portfolio. The Bank has opted for safety rather than high yields and high risks, not unlike the Alaska Permanent Fund. The return on the fund was about 15 percent in 1998 and 14 percent in 1999 (see Table 7.1).

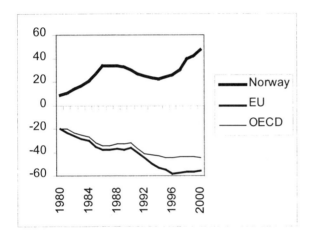

Figure 7.3. Net wealth of the public sector in percent of GDP, in Norway, the European Union, and the OECD countries. Source: *Stortingsmelding* (Report to Parliament) No. 1, 1999-2000, p. 77.

Even if the Petroleum Fund is a relatively new construct and not explicitly established to make the petroleum wealth permanent, it would be wrong to say that the oil revenues have otherwise been squandered on unsustainable consumption. The total government cash flow from oil and gas since the petroleum industry got started in 1971 has been estimated at 1100 billion 2000-kroner, of which 800 billion has been used to cover deficits on the government budget.[82] An unknown share has been used for public investment in infrastructure, education and other projects which have enhanced the productivity of the Norwegian economy, but how much and with what degree of success is unclear. Furthermore, in the fat years of high oil prices the petroleum revenues were used to pay off public debt. Figure

---

[82] Stortingsmelding (Report to Parliament) No. 1, 2000-2001, "Nasjonalbudsjettet 2001".

7.3 shows the net public debt in Norway, the European Union, and the OECD countries since 1980. The figure looks a bit like a mountain being reflected on a lake; while Norway has a substantial net public wealth the other industrialized countries are below the waterline, with a substantial public debt. In the fat years up to the oil price collapse of 1986 the Norwegian public wealth increased, and then fell until 1995 when the first deposits were made into the Petroleum Fund.

Yet another way in which the petroleum wealth has been invested is through the establishment of the state oil company Statoil. The company was from the beginning given a fat share in the oil fields developed by foreign companies, which had to pay the exploration costs for the fledgling and cashless company Statoil. This enabled Statoil to grow, and the company is now engaged in oil exploration in distant lands and, needless to say, long since able to pay its own way. There has for some time been talk of privatizing Statoil, at least in part, and to endow the company with a substantial share of the government's equity sharing arrangements in fields on the Norwegian shelf. Judging from the turn the Norwegian public debate on the oil revenues is taking as of late 2000 it could be argued that by letting Statoil and other oil companies take over the equity sharing arrangements the chances of making the petroleum wealth permanent would be much improved.

CHAPTER EIGHT
CONCLUSIONS

"In fact, if we take any cognizance of our material progress at all, we cannot help but reflect that most of this progress has taken the form of converting natural resources into more desirable forms of wealth. If man had prized natural resources above his own product, he would doubtless have remained a savage, practicing "conservation." Not only has man hastened to convert resources into equipment, but also he has invented equipment which will hasten the exploitation of resources."

Anthony Scott (1955): *Natural Resources. The Economics of Conservation.*[83]

It is time now to pull the threads together. We set out with a very broad and general view and narrowed the focus successively to problems that are more down to earth and manageable, but no less interesting for that matter. We began with some perhaps unorthodox remarks about the sustainability of human civilization in a changing world. We went on to consider an apparently unsustainable activity, the extraction of minerals for the production of energy and materials, observing that the only way of making such activity sustainable is to accumulate knowledge to ultimately do without such sources. But at the global level, that need is a long way off. We zoomed in on the individual country, state or province with a limited supply of non-renewable minerals of good quality. If that country, state or province wants to preserve its riches it will have to invest an appropriate portion of its mineral rents in renewable wealth of some kind. We looked at four cases where investment funds have been used for this purpose. Nauru's phosphate funds have largely failed, but the available information on these funds and their practices is limited. This leaves us with the petroleum funds of Alaska, Alberta and Norway.

---

[83] Page 16-17 in the 1973 edition.

Perhaps the most striking conclusion emerging from the track record of the three petroleum funds is the limited extent to which non-renewable petroleum wealth has been preserved, at any rate if we judge by the amount of money deposited into these funds. In Alaska the statutory share of the oil revenues of the state transferred to the Permanent Fund is of the order of 10 – 15 percent. In Alberta the goal was 30 percent in the beginning, but deposits to the Alberta Heritage Fund came to a halt in 1987. In recent years, two thirds to three quarters of the net cash flow of petroleum revenues accruing to the Norwegian government have been deposited into the Norwegian Petroleum Fund, but that was after nearly twenty years of oil extraction and with revenues at their all time high.

From the discussion in Chapter Three we know that it would not be necessary to direct all the petroleum rent into an investment fund in order to preserve the mineral wealth; if the return on the fund is reinvested until the rents have run out one could do with less and consume some of the mineral rents as they accrue. How much of mineral rents one could consume was shown to depend on the length of the extraction period and the rate of return on the fund. Other factors that would affect how much can be consumed are the rate of population growth and the desired time profile of consumption. In Alaska only a part of the return on the Permanent Fund has been reinvested. In Alberta the return on the financial part of the Heritage Fund was reinvested initially, but that did not last long. In Norway the return on the Petroleum Fund has been reinvested so far, and the current savings appear adequate to preserve the petroleum wealth, but the history of the Norwegian Petroleum Fund is too short for drawing any firm conclusions. Apart from these funds, some wealth may have been preserved by productive investment in infrastructure, education, etc., over and above what would have happened without the oil money.

But even if the oil funds have only been partially successful in preserving petroleum wealth they have probably meant that more has been preserved than otherwise would have been the case. At least some countries without any such funds have not saved much of their petroleum rents. In the United Kingdom total savings have declined rather than increased after the oil fields in the North Sea came on stream. In the Netherlands very little of the windfall gains of rising gas prices appears to have been saved. Norway, on the other hand, managed to save a substantial share of its petroleum rents before its petroleum fund was set up.

All three funds have not been equally successful. In the broadest of terms, the criterion of success is that oil wealth has been transformed to self-renewing wealth. When this is done by way of funds charged with making financial investments the success is easy to measure. The two main criteria are (i) has the real value of the fund been maintained, and (ii) is the rate of return acceptable? The latter question, in particular, may be further pursued, such as to considering in detail the combination of average returns and the variability of returns, but in principle these questions are easy enough to answer as long as we restrict our attention to financial investments. Whether financial investment is the best method of preserving wealth is a different question, and we have discussed at some length investments in assets that do not yield a financial return, such as education, health and infrastructure.

Given the short history of the Norwegian fund, little can be concluded from its experience. Furthermore, the five years or so since the fund began to accumulate assets have been an almost uninterrupted boom time for the Norwegian economy, and it remains to be seen how the Norwegian politicians will respond to more difficult times that are likely to come sooner or later. In fact, as of late 2000 they are having enough problems in dealing with exceptionally good times and the resulting euphoria. The two North American funds have, on the other hand, been around since 1976. While the Alaskan fund has grown steadily the Albertan fund has stagnated and even declined. The government of Alberta stopped putting money into the Heritage Fund in 1987. From then on the wealth transformation became in fact negative, as the real value of the Heritage Fund was eroded by inflation and the government of the province started to accumulate debt which for a time exceeded the assets of the fund. In terms of transforming non-renewable wealth to permanent wealth the Alaskan fund has been a success while the Albertan one has been a failure.

In terms of return the Albertan fund has also been much less successful than the Alaskan one; the return on the Alaskan fund has been high and fairly stable while the return on the Albertan fund has been lower. While the Alaskan money has been invested in stocks and bonds outside the state the Albertan money has been partly invested in public corporations in the province, some of which have failed or lost money. The Albertan fund has, however, probably been more successful than the realized rates of return suggest, as some of its money has been invested in assets with intangible benefits such as public parks, scholarships and medical research, but little research seems to have been done on the effects of this.

Success and incentive mechanisms

What, then, is the key to success, or the lack of it? Partly it lies in formal rules and regulations. The constitution of Alaska stipulates that a certain minimum of oil revenues must be deposited into the Permanent Fund, and the legislature has decided that the real value of the fund must be preserved. The deposits into the Alberta Heritage Fund were at the discretion of the provincial legislature, and in the wake of falling oil prices and government income from oil and gas it chose to use the oil money to cover current expenditure. To safeguard a process that takes a long time to complete, such as the transformation of mineral wealth, there is something to be said for clear rules that are not easily amended by a parliamentary majority. On the other hand, little would be gained if those rules can be and are circumvented, such as when a parliamentary majority passes budgets that increase public debt while adhering to formal rules of earmarking oil money into a special fund. The history of the Heritage Fund illustrates this to a degree; not only did the legislators of Alberta neglect to transform petroleum wealth, they also built up debt that for a time exceeded the assets of the previously accumulated investment fund.

But behind formal rules and parliamentary decisions lie the attitudes of the general public and the incentives faced by parliamentarians. Both are, in turn, influenced by the institutions and incentive structures in the country at hand. Therefore, to facilitate preservation of mineral wealth it is important to design rules and institutions which enhance the public interest in wealth conservation and strengthen the incentives parliamentarians have for making decisions conducive to that purpose. There are lessons to be learned, in this regard, from the history of the two North American funds. The dividend program of the Alaska Permanent Fund is an excellent example of a mechanism which creates a strong public interest in the fund. The annual dividend is a substantial enough "unearned income" for each individual and the average family that they would not be happy to see it disappear, particularly low income families. This mechanism was put to test in the 1999 referendum on whether to use some of the return on the Permanent Fund to pay for the expenditures of the state of Alaska, a proposal that was soundly defeated. Sooner or later, however, the state will have to do something about its budget deficits. It is highly likely that the voters will want to keep their dividends and have the state income tax reintroduced rather than use the return on the fund to cover the deficit. More people would gain from keeping the dividends than would lose from higher taxes, because income is not evenly distributed. But without the fund there would be no dividends. Therefore, keeping the dividend program helps preserving and building up the fund.

The Norwegian experience also supports the argument that giving individuals a direct stake in mineral rent investment funds is crucial for making sure such funds are built up and preserved. Ever since the oil price recovered from the abyss into which it fell during the Asian Crisis, the pressure for using the oil money for all the good purposes that are so easy to find has been increasing in direct proportion to the rise in the oil price and the revenue of the Norwegian government. The typical Norwegian voter is apparently not too impressed by arguments grounded in fiscal prudence but wants these riches used for satisfying immediate benefits.

The public attitude in Alberta towards the Heritage Fund stands in a marked contrast to the public attitude in Alaska to the Permanent Fund. The public attitude in Alberta is rather indifferent, and a major effort in the mid-1990s to arouse interest among the public for the question whether to preserve the fund or not was not notably successful. The reason probably is that the fund in no way directly affected the residents of the province; there was no dividend program and the income of the fund disappeared into the general expenditure account of the government. Many Albertans must have asked the question whether there was any point in preserving the fund, given that the province was accumulating debt on a substantial scale. Some arrangement such that would have tied the interests of the Albertan public to the fate of the Heritage Fund would undoubtedly have helped preserve it and might even have put pressure on the legislature to continue building it, but only if the legislature had abstained from simultaneously accumulating debt would this have amounted to a real increase in permanent wealth.

This leads on to the question does earmarking oil money for an investment fund really guarantee that transient oil wealth will be made permanent? Those who are against earmarking money in this way argue that there would be little point in earmarking certain incomes for an investment fund if the government is able to run an offsetting deficit on its general budget. On the other hand, one may ask whether an explicit commitment to channel income from non-renewable resources into a permanent fund would not constrain legislatures and governments in their borrowing to finance their general expenses. If so, that would be an argument for earmarking oil money for an investment fund, akin to the arguments for setting up autonomous institutions like independent central banks with a mandate to preserve the value of money, as this constrains politicians from actions they might otherwise be tempted to take. The evidence seems rather to support the contention that this device works with respect to independent central banks, even if it is quite possible for politicians to undo much of an independent central bank's effort by following expansionary fiscal policies.[84]

Arrangements tying the interest of the public to the financial health of a mineral rent investment fund need not just be dividend programs of the Alaskan type. Another possibility is to use the investment fund as a retirement fund, as discussed in the preceding chapter on Norway. Each generation of retirees would have an obvious interest in a fund on which their retirement payments depend. Since people in economically advanced

---

[84] On the independence of central banks and inflation, see Romer (1996), and Alesina and Summers (1993).

countries can expect to spend several years in retirement, this would ensure an ongoing interest in the fund across the generations. As noted above, this arrangement would be less harsh on people of working age than it may appear; given that there is a certain level of income to be provided to the retirees, financing this out of a return on an investment fund rather than through a pay as you go system would mean lower taxes for the working generation. This is particularly true of funds that are invested abroad, and given that we restrict attention to people of working age in the country, state or province whose inhabitants have claims on the fund. In an economy where the fund were invested domestically it would only help pay for the retirement to the extent it would increase economic growth over and above what otherwise would have happened. Without income from foreign assets, the retirement payments would have to be financed by the current income that is generated in the country where the retirees live. Making retirement payments from incomes of investment funds only means that this is done by way of capital income instead of taxes, but the source of both is the income that is currently generated by the productive forces in society.

But in order for these incentives to work, the beneficiaries of the return on investment funds accumulated from mineral rents cannot be too many, as in that case the share of each becomes so insignificant that no one cares. This raises the question of how the constituency of beneficiaries is or should be circumscribed. Sometimes history has drawn national and provincial borders in a way that channels the benefits of mineral rents to relatively few people. Alberta, Alaska and Norway are all examples of this. In other cases one can foresee resentment from the citizens at large against a "chosen few" benefiting from the mineral income. The lawsuit against giving long time Alaskans a larger dividend than more recent settlers comes to mind, and so does the case of Alberta whose oil wealth caused much resentment in the rest of Canada and gave impetus to a federal energy policy for redistributing some of this wealth to the rest of the country. In large countries with heterogeneous populations, channeling of mineral incomes to certain groups might add fuel to ethnic and regional strive, but this is not the only possible outcome; much trouble in such countries sometimes results from ethnic or regional groups which feel that their regional resources are being squandered on distant citizens with a doubtful claim on them.

Schemes where the incomes of individual citizens are not directly related to the financial health of the investment fund might also work. A fund the incomes of which are at the disposal of the parliamentary majority ought to have a certain appeal to politicians. The time horizon of parliamentarians does extend beyond the 4 – 5 year electoral cycle, if for no other reason because the ones presently in power have a certain probability to remain in power after the next election, given their aspirations to do so and their efforts to make them come true. The Norwegian Petroleum Fund is an interesting experiment in this regard. The general public has no direct stake in it, unlike in Alaska, the fund being entirely at the mercy of the parliamentary majority. How will the politicians decide to us it? Will they be able to withstand the apparent indifference and possibly outright hostility vis-á-vis the fund among the electorate? It is too early to tell.

Are there lessons to be learned for other countries or provinces from the experience of the state of Alaska and the province of Alberta? Those two, and Norway as well, are somewhat special cases, close to being "resource enclaves," where the oil revenues accruing to their governments are quite large. Alaska is the most extreme, with oil having contributed between 75 and 90 percent of the state revenue since the Prudhoe Bay field came on stream in 1977. In Alberta the oil and gas revenues have been 15 – 25 percent of the total revenue of the provincial government in recent years, and the corresponding percentage in Norway is 10 – 20. Alberta invested its money at home, either in the province or the rest of Canada, an investment policy that was not entirely successful. Both Alaska and Norway have opted for investing their funds out of state and country, respectively, owing to limited opportunities at home and a desire to diversify.

The most immediate lessons would be for other small economies where the mineral sector is large, and some of them have indeed traveled down this road. Kuwait has done so successfully, Brunei less so, and Nauru succeeded for a time with its phosphate funds and then failed. As for larger and more diversified economies, the investment fund idea would still seem a good one, as a first step towards making mineral wealth permanent. The first step is knowing what the resource rents are, and channeling them into an investment fund would make them visible. But the best use of the mineral rent investment fund need not be financial investments abroad; for a large and diversified economy, and for a large, underdeveloped country, the best opportunities may be at home. For an economically underdeveloped country the greatest contribution to development would most likely be investment in education, infrastructure, or production equipment in its own economy.

Governments, savings and institutions

Implicit in our arguments above are two conjectures, (i) governments must, and are able to, make decisions about total savings and investments, (ii) society is able to make conscious decisions about rules and institutions for a clearly defined purpose. Both may be called into question. We close the book on the perhaps pessimistic note that neither of these need in fact be satisfied.

First we address the need to have governments make decisions about total savings in the economy, particularly savings out of mineral rents. There is a theorem stating that it does not matter how a government finances its expenditure, the effect on private consumption will be the same irrespective of how much is financed through taxes and how much is financed by borrowing.[85] The argument is that consumers are forward looking and rational; they will understand that public borrowing today will mean higher taxes tomorrow, so if the government borrows today the contemporary consumer will save more today in order to be able to pay tomorrow's higher taxes. The implication for government savings from mineral rents is that consumers will adjust their private savings in accordance with government savings. If the government builds up financial wealth without making any obvious commitments to spend more at some future date the rational and forward looking consumer will conclude that the future taxes will be lower than otherwise, so that his or her savings need not be as great as otherwise. In this scenario the private sector could undo much or all of the public savings of mineral rents.

Apart from some very strong assumptions that must be satisfied for this theorem to hold, it is probably a much too stylized picture of consumer behavior to give much guidance as to what consumers will or will not do, and what governments should or should not do. There is reason to call into question the ability and even the interest of the average consumer to make predictions far into the future.[86]

If we accept that consumers are less well informed and more short sighted than governments are, there is a case for having governments make decisions about total savings in the economy, particularly about savings of mineral rents, in the hope that they will be more forward looking and better informed. Furthermore, the average consumer is not likely to be able to see through all future implications of government savings; his or her individual savings decisions are likely to be governed by more short term and egocentric considerations than ought to govern decisions affecting the entire economy. The implication of this is that savings decisions by government will not be entirely offset by contrary decisions by consumers, that governments are indeed able to influence how much is being saved out of mineral rents for example.

Governments do not, however, always act in the best public interest. It happens that they use resource rents wastefully and squander mineral wealth. The reasons why governments fail are varied. Sometimes they fail

---

[85] This is the so-called Ricardian equivalence theorem made famous by Barro (1974).

[86] An empirical investigation by Bean (1987) indicated that consumers' decisions were more "naive" and short sighted than the theorem would have us believe.

because people in government can pursue their own interests without being answerable to anyone except those who ultimately back up their power with bombs and bayonets. But even democratic governments accountable to the voters do not always spend money in the public's best interest. Catering to special interests often provides greater political gains than acting in the general interest; general interest is typically fragmented and unorganized while special interests are focussed and organized. Furthermore, acting in the interest of the future rather than the present may provide uncertain political gains. The short-sightedness of governments need not be a fault of their own, it could be a reflection of short-sightedness among the electorate; in a democracy politicians cannot depart too far from the preferences of the electorate if they are to survive as such. Seen from this angle, the democratic form of government and the accountability of politicians are constraints making it necessary to set up a system of incentives which makes the general public interested in wealth management for the long term instead of wasting mineral rents for a short term gain, a system that would save people from their own short-sightedness and narrow interests as it were. Enlightened authoritarianism might also work, and possibly work better, for ensuring that mineral wealth is preserved, but there is no guarantee that authoritarian governments will be enlightened and still less that they would remain so should they be enlightened; most authoritarian governments probably are closer to being unenlightened and self-serving.

Then there is the question whether society is able to make a conscious choice about the design of institutions, like designing a piece of machinery to do specific tasks, choosing the ones that would best serve a given purpose. There is reason to believe that the evolution of economic institutions is a lot more haphazard than that. The Alaska Permanent Fund owes much to the visions and efforts of a previous governor of Alaska, Jay Hammond, and his election was a close call twice.[87] There is no doubt that the design and purpose of the Permanent Fund was clear and carefully thought through in his mind but the reason why he got the state legislature to go along with him seems to have been more dependent on chance. His own proposal was in fact watered down quite a bit by the legislature. And why do economic institutions survive? An intriguing theory of economic institutions and their evolution is that they come about for more or less irrelevant reasons, like random mutations, but those that fulfill a purpose survive, even if they turn out to serve purposes quite different from those originally envisaged. Hence an evolutionary theory of economic institutions. The Alaska Permanent Fund survives, it may be argued, because of the dividend program. Some would argue that this is due to its populist appeal among the

---

[87] See his autobiography (Hammond, 1994).

electorate and in fact detrimental for putting the state's finances on a sound footing, which would require replacing the declining oil wealth with the return on the fund as a source of revenue to defray public expenditure in the state. However that may be, the Permanent Fund came about, and it has survived so far because many enough have felt that it serves a purpose.

No rules or structures can replace informed, forward looking decisions taken in the interest of the general public. But some rules and structures are more conducive towards such decisions than others. Democratic governments fail largely because those who elect them are swayed by narrow special interests and short-sightedness and bring their influences to bear on those whom they elect. In such environments it would appear that mineral rent investment funds have a role to play in helping to conserve mineral wealth for the future. That role consists in helping politicians to make informed and forward looking decisions with regard to the use of mineral revenues. But the funds will only be able to play that role if they are anchored among the general public, that is, if the general public perceives that such funds will provide some benefits for the individual or the household.

REFERENCES

Alesina, A. and L.H. Summers (1993): Central Bank Independence and Macroeconomic Performance. *Journal of Money, Credit and Banking*, Vol. 25, pp. 151 – 162.

Andaya, B. and L. Andaya (1982): *A History of Malaysia*. Macmillan, London.

Asian Development Bank (1998): *Report and Recommendation of the President to the Board of Directors on a Proposed Loan and Technical Assistance Grant to the Republic of Nauru for the Fiscal and Financial Reform Program*. Asian Development Bank, Manila.

------ (1999): *Country Assistance Plan (2000 – 2002). Republic of Nauru*. Asian Development Bank, Manila.

Auty, R.M. and R.F. Mikesell (1998): *Sustainable Development in Mineral Economies*. Clarendon Press, Oxford.

Barro, R. (1974): Are Government Bonds Net Wealth? *Journal of Political Economy*, Vol. 82, pp. 1095 – 1117.

Bean, C. (1987): The Impact of North Sea Oil. *In* Dornbusch and Layard (Eds.), pp. 64 – 96.

Bjerkholt, O., Ø. Olsen and J. Vislie (Eds., 1990): *Recent Modelling Approaches in Applied Energy Economics*. Chapman and Hall, Great Britain.

Cappelen, Å. and E. Gjelsvik (1990): Oil and Gas Revenues and the Norwegian Economy in Retrospect: Alternative Macroeconomic Policies. *In* Bjerkholt, O., Ø. Olsen and J. Vislie (Eds.), pp.125-152.

Chalk, N.A., M.A. El-Erian, S.J. Fennell, A.P. Kireyev, and J.F. Wilson: Kuwait: From Reconstruction to Accumulation for Future Generations. *IMF Occasional Paper* No. 150, April 1997.

Courchene, T.J. and J.R. Melvin (1980): Energy Revenues: Consequences for the Rest of Canada. *Canadian Public Policy*, Vol. 6, Supplement on the Alberta Heritage Savings Trust Fund, pp. 192 – 204.

Dornbusch, R. and R. Layard (Eds., 1987): *The Performance of the British Economy*. Clarendon Press, Oxford.

Erlandsen, H.C. (1982): *Olje*. Bedriftsøkonomens forlag, Oslo.

Forsyth, P.J. (1986): Booming Sectors and Structural Change in Australia and Britain: a Comparison, *in* Neary and Wijnbergen (Eds.), pp. 251 – 283.

Garnaut, R. and A.C. Ross (1975): Uncertainty, risk aversion and the taxing of natural resource projects. *Economic Journal*, Vol 85, pp. 272 – 287.

------- (1983): *Taxation of Mineral Rents*. Clarendon Press, Oxford..

100

Gelb, A. and associates (1988): *Oil windfalls: Blessing or Curse?* Oxford University Press, New York.

Goldsmith, S. (1998): From Oil to Assets: Managing Alaska's New Wealth. *Fiscal Policy Papers*, Institute of Social and Economic Research, University of Alaska, Anchorage, No. 10, June 1998.

----- (1999): Safe Landing: Charting a Flight Path Through the Clouds. *Fiscal Policy Papers*, Institute of Social Policy and Economic Research, University of Alaska, Anchorage, No. 12, October 1999.

Gray, L.C. (1914): Rent under the Assumption of Exhaustibility. *Quarterly Journal of Economics*, Vol. 28, pp. 466 – 489.

Gylfason, Thorvaldur (1999): *Principles of Economic Growth*. Oxford University Press, Oxford.

Hammond, J. (1994): *Tales of Alaska's Bush Rat Governor.* Epicenter Press, Fairbanks.

Hannesson, R. (1996): *Fisheries Mismanagement. The Case of the North Atlantic Cod*. Fishing News Books, Oxford.

Hotelling, H. (1931): The Economics of Exhaustible Resources. *Journal of Political Economy*, Vol. 39, pp. 137 – 175.

Jevons, W.S. (1906): *The Coal Question.* Third Edition, with a foreword by A.W. Flux. Macmillan, London.

Karl, T.L. (1997): *The Paradox of Plenty.* University of California Press, Berkeley

Kremers, J. (1986): The Dutch Disease in the Netherlands, *in* Neary and Wijnbergen (Eds.), pp. 96 – 135.

Mumey, G. and J. Ostermann (1990): Alberta Heritage Fund: Measuring Value and Achievement. *Canadian Public Policy* Vol. 16, pp. 29 – 50.

Neary, J.P. and S. van Wijnbergen (Eds., 1986): *Natural Resources and the Macroeconomy.* Blackwell, Oxford.

NOU (1983): Petroleumsvirksomhetens framtid (The Future of the Petroleum Activity), *Norges offentlige utredninger* (NOU) 1983:27.

Pratt, L.R. and A. Tupper (1980): Discretion and democratic control. *Canadian Public Policy*, Vol. 6, Supplement on The Alberta Heritage Savings Trust Fund, 1980, pp. 254 – 264.

Pretes, M. (1988): Conflict and Cooperation: The Alaska Permanent Fund, the Alberta Heritage Fund and Federalism. *American Review of Canadian Studies*, Vol. 18, pp. 39 – 49.

----- (1999): Managing Resource Wealth in North America: A Trust Fund Approach. The North American Institute, http://www.northamericaninstitute.org.

Pretes, M. and M. Robinson (1988): Beyond Boom and Bust: A Strategy for Sustainable Development in the North, *Polar Record* Vol. 25, pp. 115-20.

------ (1990): Alaskan and Canadian Trust Funds as Agents of Sustainable Development. *In* Saunders, J.O. (Ed.).

Robinson, M., M. Pretes and W. Wuttunee (1989): Investment Strategies for Northern Cash Windfalls: Learning from the Alaskan Experience. *Arctic*, Vol. 42, pp. 265 – 276.

Romer, D. (1996): *Advanced Macroeconomics*. McGraw-Hill, New York.

Sachs, J. and A. M. Warner (1995): Natural Resource Abundance and Economic Growth. *National Bureau of Economic Research*, Working Paper 5398.

Saunders, J.O. (Ed., 1990): *The Legal Challenge of Sustainable Development*. Essays from the Fourth Institute Conference on Natural Resources Law. Canadian Institute of Resources Law, Calgary.

Scarfe, B.I. and T.L. Powrie (1980): The Optimal Savings Question: An Alberta Perspective. *Canadian Public Policy*, Vol. 6, Supplement on the Alberta Heritage Savings Trust Fund, pp. 166 – 176.

Scott, A. (1973): *Natural Resources. The Economics of Conservation*. The Carleton Library No. 68, McClelland and Stewart Ltd., Toronto.

Smith, P.J. (1991): The Politics of Plenty: Investing Natural Resource Revenues in Alberta and Alaska. *Canadian Public Policy*, Vol. 17, pp. 139 – 154.

Smith, R.S. (1980): Introduction. *Canadian Public Policy*, Vol. 6, Supplement on the Alberta Heritage Savings Trust Fund, pp. 141 – 148.

Stauffer, T.R. (1988): Oil Rich: Spend or Save? How Oil Countries Have Handled the Windfall, *in* Wealth Management, A Comparison of the Alaska Permanent Fund and Other Oil-Generated Savings Accounts Around the World, *Alaska Permanent Fund, The Trustee Papers* No. 2, Juneau 1988.

Stevenson, G. (1980): Political constraints and province-building objective. *Canadian Public Policy*, Vol. 6, Supplement on the Alberta Heritage Savings Trust Fund, pp. 265 – 274.

Thomas, C.S. (Ed., 1999): *Alaska Public Policy Issues*. The Denali Press.

Warrack, A.A. (1995): Alberta Heritage Fund: Opportunity to Restructure Towards Sustainable Economic Development. *In* S.F. Zerker (Ed.): pp. 157 – 180.

Warrack, A.A. and R.R. Keddie (undated): *Alberta Heritage Fund vs. Alaska Permanent Fund: A Comparative Analysis*. Faculty of Business, University of Alberta, Edmonton. Available on the website of the Alaska Permanent Fund.

Weeramantry, C.G. (1992): *Nauru. Environmental Devastation under International Trusteeship*. Oxford University Press (Australia).

Zerker, S.F. (Ed, 1995): *Change and Impact. Essays in Canadian Social Sciences.* The Magnes Press, The Hebrew University, Jerusalem.

Websites

Alaska Permanent Fund: http://www.apfc.org

Alberta Treasury: http://www.treas.gov.ab.ca

Asian Development Bank: http://www.adb.org

Bank of Norway: http://www.norges-bank.no

Government of Norway: http://www.odin.dep.no/

Statistics Norway: http://www.ssb.no/

INDEX

Papua New Guinea, 28
peat, 1-2, 26
pensions, 83-84
Pérez, C.A., 8
petroleum, 2, 12-15, 18, 33, 40,
53, 64, 72-73, 75, 79-88, 89-92,
95
Petroleum Economist, 5n
Phillips Petroleum, 79
phosphate, 3-4, 12, 14, 42, 47-55,
89, 95
population growth, 90
    in Alaska and US, 62, 66-67
    in Alberta and Canada, 76-77
Portugal, 27
Powrie, T.L., 73n
Pratt, L.R., 74n
Pretes, M., 12n
present value, 18-19, 34-35, 43, 85
Prince Edward Island, 69, 72
principal, 53, 59-60, 72
Prudhoe Bay, 57-58, 95
Prussia, 48
public consumption, 40, 78
public debt, 42, 53, 88, 92
public expenditure, 43, 62-65, 73,
98
public finances, 42
public goods, 14
public services, 40-43, 55, 60, 63-
67

Qatar, 27-28
Queensland, 49, 69

rate of interest, 18n, 21, 33-36, 74
rate of return, 20, 24-25, 31-34,
39-40, 42, 52, 60-61, 75, 85-86,
90-91
real estate, 53, 61
referendum,

on dividend program, 65-66,
92
on Heritage fund, 78
on Permanent Fund, 58
rent, 8, 10,18
    differential, 13, 19
    entitlement to, 26-29
    investment funds, 14, 21, 31-
    46, 62, 72, 77, 81, 90, 92, 95,
    98
    market power, 13, 19
    mineral, 13, 19, 26, 29, 31-32,
    34-44, 58, 89-90, 92-98
    oil, 62, 70, 72, 77, 81, 83, 90
    phosphate, 49, 52, 55
    resource, 95-96
    scarcity, 13, 19
    taxation of, 26, 29, 38, 40
    time profile of, 20-22, 25, 34-
    37
rentier, 51-52, 62
reserve account, 60, 82
resource rent tax, 38
retirement, 83-86, 93-94
Ricardian equivalence, 96n
Robinson, M., 12n
Romer, D., 93n
Ross, A.C., 38n
royalty, 49, 57, 59
Russia, 20, 39, 57
rW rule, 36
    see also Hicksian Rule

Sachs, J., 8n
Saudi Arabia, 17, 24, 27, 79
savings rate, 33, 36-37
Scarfe, B.I., 73n
Scotland, 28
Scott, A., 1, 17, 89
Scottish National Party, 28
short-sightedness, 14, 39, 54, 97-
98